TEXT BOOK ON
NON-TIMBER FOREST PRODUCTS

Dr. M.P. Divya
Dr. C.N. Hari Prasath
Dr. R. Ravi
Dr. V. Priyanka
Dr. B. Vinothini

Tamil Nadu Agricultural University
Coimbatore - 641 003

SCIENTIFIC® PUBLISHERS
New Delhi • Jodhpur

SCIENTIFIC® PUBLISHERS
New Delhi • Jodhpur

Published by:
Jodhpur
5 A, New Pali Road, P.O. Box 91
Jodhpur - 342 001 Rajasthan (India)
E-mail: info@scientificpub.com
Website: www.scientificpubonline.com

New Delhi
4806/24, Ansari Road, Daryaganj,
New Delhi - 110 002 (India)

ISBN: 978-81-971874-4-5
eISBN: 978-81-971874-3-8

Printed in India

About the Book

This book has been designed with 15 chapters which covers the scope, importance and status of NTFPs, important Non-Timber Forest Products viz., fibre and flosses, gums and resins, tannins and dyes, essential oils, TBOs, lac and shellac, medicinal plants and drugs, spices and edible fruits, bamboos, grasses and canes, bidi leaves, dona making, animal products and mineral products. In a holistic perspective, the information furnished in this book will definitely act as ready reckoner for the faculty who teaches a course on Non-Timber Forest Products and also for the students aspiring for Scientists and Indian Forest Service.

About Authors

Dr. M.P. Divya, is borne at the Faculty of Forestry and she is the first women doctorate in the field of forestry across India and is currently positioned as the Professor, Directorate of Research, Tamil Nadu Agricultural University, Coimbatore. She is actively involved in Teaching, Research and Extension activities for the past 29 years. She has evolved suitable agroforestry models for irrigated and rainfed land use systems and developed climate resilient agroforestry models for the state of Tamil Nadu. She has obtained external funded projects from National Medicinal Plants Board, Department of Environment, Tamil Nadu Forest Department and Tripura Forest Department. She has authored 3 books and more than 65 research articles. In addition, she has presented more than 100 research papers in the National and International Conferences. She is a recipient of 6 awards for her contribution in Forestry Research. She has travelled South Korea, Srilanka, Malaysia and Singapore for attending the conferences.

Dr. C.N. Hari Prasath, is presently working as Teaching Assistant (Forestry), Forest College and Research Institute, Tamil Nadu Agricultural University, Mettupalayam. His field of specialization includes Silviculture, Carbon Sequestration, Wood Products, Natural Gums, Forest Biodiversity Assessment, Commercial Tree Farming and Yield Estimation. He has published over 50 international and national research papers including 7 books. He worked as a Junior Research Fellow in GOI funded scheme on Afforestation and Carbon Sequestration & Senior Research Fellow in ICAR-IINRG funded project on Harvesting, Processing and Value Addition of Natural Resins and Gums. He is honoured with several prestigious awards for his contribution in the field of forestry.

Dr. R. Ravi, working as Assistant Professor (Forestry) in the Department of Forest Products and Wildlife in Forest College and Research Institute, TNAU, Mettupalayam. He is working in the field of Wood and Non-Timber Forest Products and published more than 30 national and international papers in reputed journals and he has authored six textbooks. As a distinguished teacher, he handled more than 40 courses for undergraduate forestry students and more than 15 courses for postgraduate forestry students and also guided 2 PG students. He attended more than 50 national and international conferences, seminars and workshops.

Dr. V. Priyanka, is a Teaching Assistant (Forestry) working at Forest College and Research Institute, Tamil Nadu Agricultural University, Mettupalayam. She is engaged in Professional forestry teaching for the past 3 years. She has won the prestigious Thiru. S. Kondas Award for being an Outstanding student in her master's programme. She is also a DST INSPIRE fellowship awardee in her doctoral programme. She has published 4 books and more than 15 national and international papers. Her field of expertise is in natural dyes and forest products.

Dr. B. Vinothini, is a researcher and she has worked as a senior research fellow in the field of Natural dyes, TANII scheme at Forest College and Research Institute, Tamil Nadu Agricultural University, Mettupalayam. She is a gold medallist in her doctoral programme. She has published 4 books and more than 15 national and international papers. Her field of expertise is in natural dyes and Forest Products.

Preface

Non-Timber Forest Products (NTFPs) are the forest products derived from both plants and animals whose tangible values are often made to evade reaching a nation's exchequer. Throughout the human history, these products are used for a variety of purposes like food, fodder, fibre, traditional medicine, agricultural amenities, domestic materials, construction materials, etc. Many traditional beliefs and cultures are also associated with these products. Though there are few literatures and books are available, these must be organized in the form of a textbook for use by forestry graduates, forestry scientists and also by the practicing foresters.

This book has been designed with 15 chapters in which the first chapter covers scope, importance, status and livelihood significance of NTFPs and also the role of tribal co-operative societies in NTFPs. The second chapter deals with classification of fibres, fibre classification, extraction methods and flosses. The third chapter consists of main characteristics of gums and resins, kinds, important gum and resin yielding trees, tapping methods, grading and storage and the application of gums and resins. The fourth chapter covers occurrence, properties, classification, types, tannin yielding trees, extraction methods and uses of tannin. The fifth chapter contains history, status, classification, extraction techniques and application of natural dyes. The sixth chapter deals with sources, extraction methods and uses of essential oils. The seventh chapter consists of salient features of National Biofuel Policy, important TBOs, processing and extraction methods, and biofuel extraction methods. The eighth chapter covers biology of lac insect, distribution, status and strains of lac insect, lac cultivation, processing, lac products and uses. The ninth chapter contains classification of plant drugs, extraction methods, fractionation and purification methods and value addition. The tenth chapter consists of important spices and edible fruits and its uses. The eleventh chapter deals with statewise distribution of bamboos, important bamboo species, grasses and canes with uses. The twelfth chapter covers distribution, collection, processing, packaging, storage, quality classes and marketing of bidi leaves. The thirteenth chapter delas with environmental importance of leaf plates, suitable species for dona making, dona making processes and dona plates trade in India. The fourteenth chapter contains bee keeping and sericulture. The fifteenth chapter consists of mineral products.

Authors

Abbreviations

min.	minutes
ml	milli litre
h	hour
Co	Cobalt
%	per cent
ton	tonnes
NTFP	Non-Timber Forest Products
TRIFED	Tribal Cooperative Marketing Development Federation of India
LAMPS	Large-Scale Adivasi Multi-Purpose Societies
CO_2	Carbon dioxide
EOs	Essential oils
g	gram
mm	millimeter
m	metre
cm	centi metre
sq.cm	square centimetre
TBOs	Tree Borne Oil Seeds
° F	degrees Fahrenheit
WEFs	Wild edible fruits
ISFR	India State of Forest Report
FSI	Forest Survey of India
g/m^2	Gram per metre square
FAO	Food and Agriculture Organization
m ha	million hectare
sq.km	square kilometre
IAA	Indole Acetic Acid
DM	drymatter
dfl	Disease free layings
m	metre
CSR&TI	Central Sericultural Research and Training Institute
CSTRI	Central Silk Technological Research Institute

List of Figures

List of Tables

Table of Contents

CHAPTER 1
Introduction

M.P. Divya and C.N. Hari Prasath
Tamil Nadu Agricultural University, Coimbatore

Non-Timber Forest Products (NTFP) are defined as ***"Products or Services" other than timber that is produced in the forests.*** *It can also be defined as* "Goods of biological origin other than timber derived from forests, other wooded land and trees outside forests". It is also called as Alternative and Secondary Forest Products obtained from the forests. The NTFP include Gums and Resins, Fibre and Flosses, Tans and Dyes, Bamboos, Grasses and Canes, Essential oils, Tree Borne Oil Seeds, Drugs, Spices and Edible fruits, Medicinal Plants, Bidi leaves, Animal products, Mineral Products, etc.

Classification of Non-Timber Forest Products

Non-Timber Forest Products are classified into 3 types. There are (a) NTFP of Plant Origin, (b) NTFP of Animal Origin and (c) NTFP of Mineral Origin

a) NTFP from Plant Origin

i) Gum and Resins
ii) Tans and Dyes
iii) Fibre and Flosses
iv) Essential oils
v) Tree borne Oil Seeds
vi) Edible Fruits
vii) Spices
viii) Medicinal and Aromatic Plants
ix) Bamboos and Canes
x) Fodder and Grasses
xi) Bidi Leaves

b) NTFP of Animal Origin

i) Honey and Bee Wax
ii) Lac and Shellac
iii) Tasar and other silks
iv) Shells

c) NTFP of Mineral Origin

i) Metallic minerals
ii) Non-Metallic minerals

Scope and importance of NTFP

Non-Timber Forest Products (NTFP) are important to the livelihood of millions of people in rural areas all over the world. It is generally known that NTFP have a variety of roles in maintaining human health. Many of the world's poor people rely on NTFP for food, shelter, medicines, textiles, energy and cultural artifacts. NTFP provide a source of income for many households, either as a supplement to other sources of income or as the primary source of income. India's forests are endowed with vast natural resources including non-timber forest products, such as a large variety of medicinal and aromatic plants (MAP), leaves, fruits, seeds, resins, gum, bamboos and canes. More than 3000 plant species produce economically useful NTFP in India viz., fruits and nuts, vegetables, medicinal products, gum and resins, essential oil, bamboo, rattans and palms, fibres and flosses, grasses, leaves, seeds, mushrooms, honey, lac, etc.

NTFP provide diverse employment and economic opportunities to support rural livelihoods and protect the environment. Many of the small scale forest based enterprises rely on NTFP which provide upto half the income of about 25 percent of India's rural people. Nearly half of the country's forest revenue and 70 percent of export forest revenue obtained from NTFP. The export potential of NTFP is growing as the scope of globalization increases and recognition of the health benefits of herbs become more widespread. Large quantities of NTFP are available in almost all the states of India. Hence, there is a wide scope in exploring the valuable NTFP which would increase the national economy and also create huge employment to rural poor and also forest dwellers. NTFP provide green social security to billions of people in the form of food supplements, traditional remedies, fuel and fodder and low cost building materials as well as a source of employment and revenue generation. In other circumstances, NTFP profits are the only source of income, increasing people's reliance on commercially valuable NTFP resources.

Status of NTFP

Non-Timber Forest Products (NTFP) deserve special mention because of their vast potential to support an economic development consistent with the principles of Sustainable Forest Management (SFM). It covers a wide range of products (goods & services) from thatching materials to medicinal plants. These products are the essential needs of local communities. Some NTFP such as latex, gums, resins, essential oils, spices, condiments and aroma chemicals help to promote value added processing, niche marketing and export trade. Nearly 400 million people living in and around forests depend on NTFP for their sustenance and supplemental income.

It provides as much as 50 percent of the income to around 30 percent of rural people. In this context, there is a growing realization that NTFP have a tremendous potential in meeting the subsistence needs of local people on the one hand while alleviating their extreme poverty on the other. Another striking feature of NTFP is that they can be used on a sustainable basis under scientific manner without adversely affecting the resource base.

The contribution of NTFP to the forestry sector in most countries is significant. A recent valuation undertaken by the Ministry of Environment and Forests in India estimated that 220 million tonnes of fuelwood, 250 million tonnes of grass and green fodder and 12 million m^3 of timber are removed from India's forests annually. These products are estimated to be worth US$ 10 billion. In India, NTFP provide about 52 percent of total forest revenues and 61 percent of forest based employment. Nearly 1.3 billion people living in and around forests in India rely on NTFP as a critical component for their sustenance. In Madhya Pradesh, the NTFP which are primarily collected by tribal women are worth of more than Rs.24 billion annually. The income from NTFP ranged from Rs.667 to Rs.9856 per hectare per year with a mean of Rs 5260. Furthermore, revenues from NTFP have been growing faster than revenues from timber in the past. For example, compound growth rates in revenue from NTFP in India during the 2013/14 to 2020/21 period were 43 percent higher than those for timber. Export earnings from NTFP on average account for about 60 to 70 percent of total export earnings from forest products and this proportion has been rising.

There is a considerable scope for increasing exports by exploiting untapped resources as the current production of most NTFP is estimated to be about 60 percent of the potential production. In case of non-edible fibres and flowers, production is only 7 and 12 percent, respectively, of the potential production. It is felt increasingly that management and development of NTFP resources is essential for various reasons. First, forest management focused on the production of NTFP may be ecologically and economically sustainable provided that extraction rates do not exceed the maximum sustainable yield. Tribal communities have been involved in NTFP utilization for centuries without destroying the resource base. Managing forests for the production of NTFP also implies maintaining biological diversity of both plant and animal species. In addition, non-timber forest products are the vital source of livelihood for a large proportion of the poor living in or close to the forest in most tropical countries. In West Midnapore district in West Bengal, many village communities derive as much as 17 percent of their annual household incomes from NTFP.

It is also estimated that upto 35 percent of the income of tribal households in India comes from the collection of unprocessed NTFP. Also, since NTFP involve a large variety of seasonal products, returns are frequent and relatively continuous. Moreover, local processing of NTFP can increase off-farm rural employment opportunities. Small scale forest based enterprises which depend on NTFP, provide upto 60 percent of income for 30 to 38 percent of the rural labour force in India. NTFP also provide food security to large low income populations, their cattle and other domestic animals, particularly during droughts or famines. The tribal people are living near and inside the forests. The rights and concession enjoying by them should be fully protected. Having regards to the symbiotic relation between the tribal people and forests, a primary task of all agencies responsible for forest management, including forest development corporations should be to associate the tribal people closely in protection, regeneration and development of forests as well as to provide gainful employment to people living in and around the forests.

Livelihood significance of NTFP

In India, NTFP are linked to the socio-economic and cultural activities of forest dependent communities in the area in a variety of ecological and geoclimatic conditions in different parts of the nation. Tribal livelihood systems range significantly between areas and ethnic groups and are influenced by ecological, historical and cultural variables. Since time immemorial, tribal communities have generally occupied forest regions, living in isolation from mainstream society while maintaining harmony and a symbiotic relationship with nature. The percentage of NTFP collected by communities for subsistence purposes varies from state to state which ranged from 5.40 to 55 percent across the country. In Manipur, approximately 90% of the populations rely on forest goods as a primary source of income with 250000 women involved in the collection of forest products. Around 75% of forest dependent inhabitants in Chhattisgarh supplement their sustenance with tubers, flowers and fruits throughout the year.

In addition, the collection and processing of economically valuable NTFP species creates at least 35 million man days of employment. NTFP are projected to be used by 275 million impoverished rural people in India or 27% of the total population for their subsistence and commercial income. In India, around 70 percent of NTFP collection occurs in the tribal belt, while accounts for 55 percent employment opportunities for tribal communities. While NTFP gathering is an important source of income and employment for forest dwellers, downstream processing and trading operations have a multi-fold impact on the economy. Due to an increase in demand for natural products, NTFP have gained significant commercial significance in trade and industry. Trade in NTFP can operate as a financial incentive for forest conservation by offering a source of income from resources that would otherwise be considered low value. Furthermore, in this era of globalization, millions of invisible hands from forest collectors to food and medication firms in different parts of the world are determining the marketing and pricing of NTFP.

Challenges and Emerging Issues

NTFP have attracted new attention in international debates due to their potential role in livelihood and poverty reduction, making it critical for governments to implement pro-poor reforms in the forest sector to maintain and enhance the livelihood advantages that forests bring to the poor. NTFP laws and policies can enhance ecological sustainability, trade justice and improved rural livelihoods with more information, effective stakeholder engagements and strategic policy making processes. Policies and rules should reflect the amount of commercialization as well as the diversity of NTFP resources, markets and stakeholders.

Incentives and supportive legal frameworks such as government assistance for the producer, trade and processing organizations, market access and premium prices through certification, tax breaks and outreach and education on new policies and regulations are all beneficial to NTFP policies. A more extensive regulatory structure including permits, quotas, tariffs and trade restrictions may be required in some circumstances, particularly when there is a rapid and substantial commercial demand. The central and state governments have been running several programs to support the tribal communities in increasing their earnings from the NTFP, with the latest being the Van Dhan Yojana which aims to ensure the fair return of the NTFP. Tribal leaders, environmentalists and experts working with the tribal communities believe a continuous focused effort is required to ensure that the tribal people get a proper value of the NTFP. They also believe that it also requires a special focus on women of the tribal communities.

The population of the tribal community in India is 104.5 million (10.45 crore), which is about 8.6 percent of the total population of the country. Of that, about 90 percent live in rural areas where apart from farm work, they are primarily dependent on the NTFP for their livelihood. Though authorities emphasize that the tribal people are the real owners of the NTFP and a minimum support price (MSP) is announced for several NTFP, the ownership of the tribal communities largely remains notional. Those working with tribal communities argue that the MSP scheme needs to evolve beyond a mere safety net for tribals and it needs to be a better system for them to realize the true value of the NTFP. Under the MSP scheme, the government looks at fixing and declaring MSP for the selected

NTFP and then procurement and marketing of those products are undertaken by state agencies. Under this scheme, the government has spent about Rs.530 million between 2014-15 and 2018-19. In April 2018, Prime Minister Narendra Modi launched Pradhan Mantri Van Dhan Yojana (PMVDY) and stated that 'Van Dhan, Jan Dhan, and Govardan' would be the basis for transforming the rural and tribal economy in the future. This scheme is in addition to the government's efforts aimed at ensuring fair returns to the tribal community for the NTFP through the MSP scheme. Under the *Van Dhan Yojana*, the government aims to bridge gaps like the shrinking of the area under the NTFP due to the timber first policy.

Role of Tribal Co-Operative Societies in NTFP

i) TRIFED

The Tribal Cooperative Marketing Development Federation of India (TRIFED) was established in 1987 under the Multi-State Cooperative Societies Act, 1984. This body works for the social and economic development of the tribal people of the country. The Tribal Cooperative Marketing Development Federation of India (TRIFED) has launched "Sankalp Se Siddhi"- Village and Digital Connect Drive under the Ministry of Tribal Affairs. The aim of the drive is to activate the Van Dhan Vikas Kendras (VDVKs) in the villages. It has been registered as a National level Cooperative body by the Government of India. Its focus is on institutionalizing the trade of NTFP and Surplus Agricultural Produce (SAP) collected/ cultivated by them. TRIFED helps tribal people to manufacture products for national and international markets on a sustainable basis. It also supports the formation of Self Help Groups and imparts training to them.

Objectives

- Major part of the income of tribal people is dependent on the manufacturing and production of tribal products. The main objective of TRIFED is to promote marketing development for the tribal people
- It performs the role of both, a market developer and a service provider
- They support the tribal people with tools and training which can help them boost their market and production
- TRIFED helps in discovering a systematic and scientific approach for their development
- The ultimate aim is to help the different tribes of the country grow socially and economically. These people are entirely dependent on their handicrafts for a sustainable livelihood

Other Initiatives by TRIFED

In line with the socio and economic development of the tribal population of the country, TRIFED has even been a part of a few other initiatives:

- Empanelment of Tribal Producer/Tribal Artisans
- Sourcing products through its impanelled tribal suppliers. These suppliers may be individual tribal artisans, tribal Self Help Groups, agencies or NGOs working for the development of tribal people
- Pradhan Mantri Van Dhan Yojana (PMVDY)
- Van Dhan Karyakram. This has been initiated by the Ministry of Tribal Affairs. To set 50,000 Van Dhan Vikas Kendras (VDVK) in the tribal areas across the country to provide livelihood and empowerment to the tribal people.

a) Village and Digital Connect Initiative

To ensure that existing schemes and initiatives reach the tribals, TRIFED's regional officials across the country have been visiting the identified villages with a significant tribal population.

b) Tech for tribals

Tech for Tribals program aims to transform 5 crore Tribal Entrepreneurs by capacity building and imparting entrepreneurship skills to tribal forest produce gatherers enrolled under the Pradhan Mantri Van Dhan Yojana (PMVDY). The Van Dhan Vikas Yojana is an initiative of the Ministry of Tribal Affairs and TRIFED. It was launched to improve tribal income through the value addition of tribal products. The program will ensure a higher success rate of the Tribal Entrepreneurs by enabling and empowering them to run their business with marketable products with quality certifications.

c) Schemes for ensuring fair prices to tribals

Schemes such as mechanism for marketing of Minor Forest Produce (MFP) through Minimum Support Price (MSP) and Development of Value Chain for NTFP and provide MSP to gatherers of forest produces. The schemes ensure the sustainability of the resource base by addressing the problems that tribals face such as perishable nature of the produce, lack of holding capacity, lack of marketing infrastructure, exploitation by middlemen, etc.

d) TRIFOOD Scheme

It is a joint initiative of the Ministry of Food Processing Industry, Ministry of Tribal Affairs and TRIFED and it promotes value addition to NTFP. Conclusively, the main objective of the body is to empower the Indian tribal population, help them provide a livelihood and capacity building through sensitization.

ii) LAMPS

The Large-Scale Adivasi Multi-Purpose Societies (LAMPS) are the Co-operative Societies set up by the government for integrated tribal development in regions with essential tribal populations. They are generally one per taluk and their membership is controlled by adult members of any Scheduled Tribe community. The objectives of LAMPS are as follows.

- Encouragement of subsidy cum loan production plan in the field of agriculture, horticulture, animal husbandry, irrigation, forestry, cottage and rural community industries.
- Liberating the tribals from the control of money lenders through the supply of production and consumption credit.
- Purchase additional agricultural produce and forest produce from tribals and supply of necessary commodities and agricultural inputs to people to tribal.

The main action of the LAMPS should be the obtainment and providing NTFP. The major NTFP collected by LAMPS are honey, tamarind, amla or indian gooseberry, tree mass, soap nut, cinnamon, gum, grasstick, etc. Collection and marketing of NTFP is the important wage creating action taken by LAMPS.

Main functions of LAMPS

- To facilitate, co-ordinate and promote the marketing of the tribal products by its members.
- To assist in the capability and capacity building of the members relating to the marketing development of the tribal products
- To enlarge and extend the markets for tribal products through marketing development and promotion programs
- To strive for the development of a fair trade mechanism for tribal products and marketing thereof.
- To develop quality specifications and a quality preservation system for tribal products in collaboration with all related organizations, departments and institutions.

- To promote the setting up of units for production and value addition of tribal products
- To start and promote the creation of technologies and infrastructure for basic value addition on tribal products.

Suggested Readings

Dhiman KR Bawa R Raina KK and Dohroo NP 2013 *Non Timber Forest Products: Conservation Utilization and Management,* Dehradun: BSMPS Publishers & Distributors of Scientific Books

Krishna Murthy T 2010 *Minor Forest Products of India (Non-Timber Forest Products of India),* Hyderabad: BS Publications/BSP Books

Mishra TK Banerjee SK and Pal DC 2004 *An Omnibus of Non Timber Forest Products of India,* Dehradun: International Book Distributor

Smith-Hall and Carsten 2022 *The bioeconomy and non-timber forest products,* Boca Raton, Florida: CRC Press

Tewari DD 2015 *Managing Non-Timber Forest Products (NTFPs) as an Economic Resource,* New Delhi: Write and Print Publications

CHAPTER 2
Fibres and Flosses

M.P. Divya and R. Ravi
Tamil Nadu Agricultural University, Coimbatore

FIBRES

Fibres are hair like materials that are continuous ""filaments"" or are in discrete elongated pieces, similar to pieces of thread. They can be used as a component of composite materials. They can also be matted into sheets to make product such as paper or felt. Forests have been traditional source for supply of fibres and flosses. The most common examples are ropes, cordages, mattresses made by the rural people for several domestic purposes traditionally. Fibres of both vegetable and animal origin have been used quite long time for the spinning of the thread and cordage and weaving of coarse fabrics. Fibres generally occur as sclerenchyma cells and serve to provide rigidity to the plant. They are long cell with thick walls and small cavities. They are found in various parts of the plant such as stems, leaves, roots, fruits and seeds.

Classification of Fibres

Classification of the fibres is done on the basis of their origin, texture, morphology and uses. On the basis of origin, fibres can be classified into: *i) Fibres from roots, ii) Fibres from stem and iii) Fibres from leaves.* On the basis of texture, fibres can be classified into i) Soft fibre (fibre obtained from the innermost bark of the stem are called soft fibres, example: jute, hemp, flax, etc.), ii) Hard fibres (fibre obtained from leaves of plants are hard fibres, e.g., Manila hemp, etc.) and iii) Surface fibres (fibre obtained from the surface of plant parts viz., stem, leaf, seed, etc. are generally called surface fibres). On the basis of morphology, fibres can be classified into i) Hair borne on the seeds or the inner wall of the fruits, example: cotton and kapok, ii) fibres occurring in the innermost tissue or bark of the stem, example.: flax, hemp and jute, iii) fibres obtained from the leaves of the plant and which are a part of the fibres of vascular system of the leaves, example: cordage fibres, manila hemp and iv) woody fibres of trees and plants consist of various elements which constitute the fibre vascular tissues of wood and commonly been used in paper making (Fig. 2.1a & Fig. 2.1b).

Fig. 2.1a Hard fiber

Fig. 2.1b Soft fiber

a) Classification of fibres based on origin

i) Fibre from roots

Butea (*Butea monosperma*): Young roots of the plants yield fibre, which is used in making ropes and cordage. For obtaining the fibres, the young roots are dug, taken out and cut into 0.5 m long pieces. When these are green, the end portion of the root is crushed against stones, which forces the fibres to release from the root. These fibres are then caught and torn from the remaining portion of the root and removed by beating them lightly. The fibres are generally used for making ropes and cordages. The ropes are generally coarser and better for use under wet conditions. During dry period, the rope becomes brittle and breaks easily.

Pandanus *(Pandanus odoratissimus)* is another plant which is exploited for fibre from its roots. It is a densely branched shrub, found along the coastal regions and in Andaman Islands. The roots of the plant are fibrous and are used by basket makers for binding purposes. This fibre is used in making paint brushes and brushes for white wash. It is used as a substitute of bristles in brushes.

ii) Fibres from stems/bark

In some plants, the stem yields fibre. In such plants, fibres are extracted from the bast tissue of the stem of woody species. The type of the fibre varies from plant to plant. Some plants yield long and strong fibres, while some plants yield silky fibres, which are used in textiles. Retting is the commonly used process of extracting plant fibre in which the stems are submerged in ponds or streams for 10-15 days. It is the process of separating the fibres after loosening them from the woody plant tissue. This process depends on the age of plant, species, type of fibre required and the temperature of water used during processing. Over retting is harmful for the plant tissues, as they lose the luster on over treatment. The common fibre yielding species are given below.

Calotropis spp. (*Calotropis gigantea* and *Calotropis procera*) yield stem fibres. These are moderate sized shrubs commonly found in the dry tracts of the country on the wasteland or as an associate in dry mixed deciduous forests. The bark of the stem yield a white, silky, strong and durable fibre which is superior to cotton in tensile strength. It is extensively used for making fishing nets and lines, bow strings, twines, etc. Extraction of the fibre from the plant is done by steaming of the stem followed by pressing it inbetween wooden rollers and water retting followed with mallet beating. It can also be extracted by soaking the stem in water for 24-48 h followed by autoclaving to avoid encrustations. The separated fibres can be bleached to yield white flexible, cylindrical fibres. These fibres can be spun alone as well as mixed with cotton.

Marsdenia (*Marsdenia tenacissima*) is a large twinning shrub commonly found in the sub-Himalayan tracts of the country. It is occasionally found in the Khair forests and less frequently in sal forests. This fibre is used in making fishing nets and lines, bow strings, netting and cordage.

Camel's foot creeper *(Bauhinia vahlii)* is a large evergreen climber, generally found in the moist localities and shady hill slopes with rock outcrops. It is common in almost all forest types in India. It causes great damage to the forest trees. The fibre obtained from the inner bark of the plant is strong and is used for making rough ropes. The barks of *Bauhinia racemosa* also yield fibre.

White bark acacia (*Acacia leucopholea*) is a small to moderate sized tree, found in the tropical dry deciduous forests in arid and semi-arid regions of the country. The bark of the plant yields a strong fibre when soaked in water and beaten. This fibre is commonly used for making fishing nets and coarse cordage.

Flame of the forest (*Butea monosperma*), the bark of this tree also yields fibre (besides roots) which is used in making ropes and cordage. The fibre obtained from the bark is inferior to the fibre from roots. *Butea parviflora* is another species which is a woody climber, found in most parts of the country which also yields fibre from the inner part of the bark is used for making ropes. The fibre is strong and durable.

Anjan (*Hardwickia binata*) is a moderate sized deciduous tree, very common in Madhya Pradesh. It is characteristic of dry teak and degraded dry deciduous forests of Central and Southern India. The young shoots and branches of the tree yield a strong, reddish brown colour fibre. The fibre is widely used in making well ropes and also for other agricultural purposes.

Ujjain Desmodium tree (*Ougenia oojeinensis*) is a medium sized deciduous tree generally found in tropical deciduous forests. In Uttarakhand, this tree is found on rocky habitats in warm valleys. The new branches yield fibre, which is used for making ropes and cordage.

Sacred fig tree (*Ficus benghalensis*) is a large sized evergreen tree, found almost throughout the country in the plains upto 600m and in the hills. The bark and the aerial roots of the tree yield fibre, which is used as coarse rope. *Ficus cunia* and *Ficus religiosa* yield stem fibre, generally used for making ropes.

Indian screw tree (*Helicteres isora*), a moderate sized tree is commonly found in the dry and moist mixed deciduous forests of India. The fibre is extracted from the inner bark of the plant by retting it in water, pool, ponds and rivers. Rotted branches are then taken out from the water and beaten lightly to separate the fibre. The fibre is light brown to greyish green in colour, soft, silky and lustrous. It is durable, but is inferior to jute fibre in strength. The fibre is mostly used as cordage for sewing sacks, gunny bags and cattle harness. The quality of the fibre which can be spun into good yarn and woven into canvas and better quality of durable fabrics by selecting the equal aged stalks and retting them properly and scratching the dried fibre on appropriate modern machines. In Southern India (Kerala, Tamil Nadu and Karnataka), the shrub is collected from the forest for extracting fibre. Its fibres are twice durable than jute which is cultivated extensively in Kerala and West Bengal.

Dhaman (*Grewia elastica*) is a tree with grey bark that grows in sal and mixed forests in outer Himalayas. This moderate size tree is mostly cultivated in the lower hills of Uttarakhand by the farmers around their farmlands. This is a multipurpose tree which yields quality fodder, fuelwood and fibre. The bark of the stem yields strong white fibre which is used for cordages and rope making. *Grewia tiliaefolia* (Daman), *Grewia optiva* (Bhimal) and *Grewia vestita* are other species which yields a coarse, strong and yellow brown colour fibre useful for rope making and for domestic purposes. It is a small medium sized deciduous tree, 8-10 m height. The species is found abundantly in the entire Himalayan region upto 1800m. The bark of the species yields fibre. The fibre is used for making ropes and clothing.

Stinging Nettle (*Utrica dioca*) is a herbaceous perennial plant. They grow like weeds and are generally eradicated. The fibre is removed from the plants by retting. Nettle fibers are better than those of hemp and can even be used for cloth making. Nettle may also be used as a dye, producing yellow colour from roots or yellowish green from leaves.

Indian hemp (*Cannabis sativa*) is a plant native to Asia. The term hemp may refer to the tall, coarse plant itself or to the coarse tough fibre from the plant which can be used to make rope, fabrics, etc. The fiber is one of the valuable parts of the plant. It is commonly called bast hemp fibre which varies from 2.5 to 3.5 m long depending on the height of the plant.

Udal (*Sterculia villosa*) is a small to moderate sized tree, found in the tropical deciduous forests throughout the greater part of India. A coarse and strong fibre is obtained from the inner layers of the bark. The fibre strips off in broad flakes having a net like appearance. The fibre is commonly used for making bands for elephants used for dragging timber. The fibres are made into ropes for tying cattle and sheep and for making cordages. The other species belongs to the same genus are *Sterculia urens, Sterculia foetida* and *Sterculia colorata.*

Indian charcoal tree (*Trema orientalis*) is a large shrub or a small sized tree found in tropical deciduous forests. The tree yields a long light brown fibre. The fibre can be easily stripped off from the bark of the tree in thin narrow bands. The fibres, on drying, can be easily separated. The fibre is mostly used for making ropes, twine and coarse cloth.

Nilgiri nettle (*Girardinia heterophylla*) is a medium sized shrub commonly known as stringing nettle and commercially as Nilgiri nettle. It is found in the Himalayas and in the peninsular region of India. It yields a fine, soft and silky fibre, which is used for making ropes, twine and coarse cloth. The plant is not given high priority for fibre because of the difficulties like cleaning the fibre.

Pola (*Kydia calycina*) is a large sized tree generally found in tropical moist deciduous forests. It yields a strong fibre which is used for making ropes and cordage. It is also used for making drag ropes. This tree is also found in Bhabar area of the Uttarakhand.

Wild guava (*Careya arborea*) is a moderate sized tree, found as a common associate of moist mixed deciduous forests. The inner bark of the tree yields fibre which is used for making coarse cordage and ropes for tying carts and rafters. The fibre is also used in paper industries. It is a deciduous tree that grows up to 15m. Its leaves turn red in the cold season. Flowers are yellow or white in colour that become large green berries. The tree is a common associate of sal forest and grows throughout India. In Uttarakhand, it is commonly found in bhabhar and tarai regions.

Sea mango (*Cerbera manghas*) is an evergreen tree or large shrub with milky latex. It is found in the tidal forests of India commonly in the Andamans. The bark of the tree yields a fibre of minor importance.

Indian cherry or pink pearl (*Cordia dichotoma*) is a small or medium sized tree with crooked trunk. It is found in the warmer regions of the country. The bark of the tree yields fibre which is used for caulking of boats. Another species, *Cordia rothii*, *Cordia myxa* and *Cordia oblique* also yield a fibre used for caulking boats and making ropes.

Corky coral tree (*Erythrina suberosa*) is a medium sized tree, found scattered throughout the dry forests of India. The bark of the tree yields a fibre used for making cordages. A medium size tree with very attractive maroon reddish coloured flower clusters. Young stems are prickly. The fibre is obtained from bark and used for cordages. It is found grown extensively in bhabhar and tarai regions of Uttarakhand.

Indian coral tree (*Erythrina variegata*) is found in Andaman and eastern coastal states of India. The fibre is obtained from the bark and used for cordages.

Indian ash tree (*Lannea coromandelica*) is a moderate to large sized deciduous tree, distributed throughout the greater part of India upto 1000m elevation. The bark of the tree yields a strong fibre used for ropes and cordages.

Velvety miliusa (*Miliusa velutina*) is a middle sized tree, found in the moist deciduous forests. The bark of the stem is locally used for extracting fibre which is used for making ropes and cordages.

Portia tree (*Thespesia populnea*) is a compact quick growing evergreen tree, found in the coastal regions of India, commonly in the Andamans. The bark of the tree yields a strong white fibre used for cordage, fishing lines and coffee bags. It is also used for caulking boats.

Caesar weed or Congo jute (*Urena lobata*) is a shrub occurring throughout the hotter parts of India. The plant is often cultivated as a fibre crop. To obtain good quality fibre, the harvesting is suggested when the plant is full in flowers. If harvesting is done before the time recommended, the fibre obtained is less in quantity and fine in texture. If the harvesting is done after the time of flowering, the fibre obtained is less shiny and coarse in texture. Its fibre closely resembles with jute in chemical composition. It is mostly used for making ropes, sacks, cordages and carpets. Mixed with other fibres, it can be worked on machinery and can be utilised for making artificial silks, upholstery and sail cloth. *Urena repanda* is another species which yields fibre of the same quality as of *Urena lobata*.

iii) Fibres from leaves

Agave or sisal (*Agave* spp.) is the most important and valuable among the fibre yielding plants. Agave consists of about 300 species. These species are native of tropical and subtropical North and South America. They have been introduced to a large number of countries in East Africa, West Indies, Indonesia, Israel, South Africa, Philippines, etc. Agaves have been introduced in India as hedge plants. These are generally planted in gardens, fields and along the road and rail sides. The common species planted in India are *Agave sisalana, Agave fourcroydes, Agave cantala, Agave wightii* and *Agave americana.* In Uttarakhand, efforts are made to cultivate this plant in wastelands on a large scale. The planting of Agave species is done in the forest and the modern practice is the use of bulbils for plantation. The leaves of this plant yield a valuable fibre and the fibre is considered to be one of the important hard fibres. It is mostly used for making ropes, cordage and twine.

Sometimes the short fibres of the genus are used for making mops and brushes. The fibre compares favourably in durability with manila hemp for marine cordage. Recently, the fibre is used for manufacturing coarse fabrics. It is extensively used for binders, twine, fishing nets, hammocks, door mats, rugs, carpets, etc. Generally, the mature leaves are harvested twice a year, first in the month of january and secondly in the month of september. About 10-15 tonnes of leaves per ha are generally available in the plantation areas. There are several species of *Agave* which are mainly grown for fibre production. *Agave americana* is grown as an ornamental garden tree and also grown as a hedge plant on the boundaries of the field. The leaves of this plant yield about 5 percent of fibres used for making ropes and cordage. *Agave angustifolia* is commonly found in the sub-Himalayan tract and outer Himalayas. The leaves of the species yield somewhat shorter fibres. *Agave sisalana* has been planted on a large scale in several states of India. The species grows well in areas with an annual rainfall of about 700-1300 mm. It succumbs to water logging and cannot tolerate very low temperatures. *Agave vera-cruz* is native of Mexico and is also found throughout the greater part of India. It is found commonly in Assam, Bengal, Bihar and southern parts of India. The leaves yield about 1.5-2.5 per cent fibre. The fibre is coarser and stronger than the Cantala fibre and is used for making ropes, mats and cordage.

Indian sago-palm (*Caryota urens*) is one of the most important leaf fibre producing plant. This palm is found in the moist regions of western and eastern coasts in cool and shady places. The leaves of this plant yield strong fibres. The fibres are obtainable from the base of leaf sheath, petiole and flowering stock. The fibres are straight, strong and smooth and very elastic. The fibres are like horse-hair type in appearance and used in making rope of great strength. The fibres are used by fishermen for preparing fishing nets and fishing lines. They are also used for making soft brooms.

Pandanus or fragrant screwpine (*Pandanus odoratissimus*) is a densely branched shrub, found along the coastal region of India and in Andaman Islands. The leaves of the plant yield fibre used for making sacks for coffee, grains, sugar, etc. The fibre is used for making ropes, mats, cordage, hats, baskets and other fancy articles. To get the superior quality of mats, young and freshly harvested leaves are used.

iv) Fibre from grasses

Important grasses which yield fibres for making mats, ropes, cordage, fishing nets include *Eulaliopsis binata*, *Desmostachya bipinnata*, *Saccharum bengalense*, *Saccharum spontaneous*, *Themeda arundinacea*, *Phragmites* spp, *Arundo donax*, *Typha elephantina*, etc. are used for making ropes, cots and fishing nets. These are also used for making cordage and mats. Some of these grasses are highly suitable for making paper. *Eulaliopsis binata* is cultivated on a large scale in shiwalik region of Himalayas. It is also grown in Madhya Pradesh and Orissa. The grass is used for making ropes. It is also used for making paper (Fig. 2.2).

Fig. 2.2 Grass yielding fibre

Fibre Extraction Methods

Retting is a process employing the action of microorganisms and moisture on plants to dissolve or rot away much of the cellular tissues and pectins surrounding bast-fibre bundles, and so facilitating separation of the fibre from the stem. It is used in the production of linen from flax stalks and coir from coconut husks. There are two main methods that have been applied to extract bast fibres (mechanical extraction and retting). The mechanical extraction provides a simple but rapid process with high quantity of fibre yield when compared to the other retting processes. Mechanical extraction uses mechanical forces to tear bonds between the fibre and its core. Plant stalks are fed into a decorticator to break into pieces via compressive, shear, and/or impact forces. Different types of decorticators have been used for fibre processing such as the crushing roller, hammer mill, ball mill and drop weight. On the contrary, post decortication cleaning process separates the detached fibres from the mixture of fibre/core-bounded components and fine particles. However, it is hard to control the mechanical forces applied. A wide range of fibre length and high in cost are the disadvantages of the mechanical extraction.

Retting process of fibre extraction

Fibre retting is a complex process and its properties are highly dependent on the type of retting and its parameters. Under-retted and over-retted fibres made inefficient fibre separation and fibre weakening, respectively. During the retting process, phloem-derived fibre bundles are loosened from hemicellulose, lignin and pectin. Leftover fibres are rich in cellulose content and hence performing in high strength properties. Retting is a biological process, which removes non-cellulosic materials attached on the fibre bundle by enzymatic activities, consequently yielding detached cellulosic fibres. All retting processes except chemical retting use enzymatic activities to extract fibres from bundles. Water retting and dew retting apply anaerobic bacteria fermentation and the fungi colonization method, respectively, on fibre bundles, to produce enzymes that hydrolyse fibre-binding components. *Clostridium* sp. that lives in lakes, rivers and ponds is an anaerobic bacterium. It rets bast fibres by producing pectinase enzymes during the water retting process. However, anaerobic fermentation has resulted in severe water pollution, contaminated wastewater and putrid odour and hence resulted in introducing the dew retting process. Dew retting uses colonization of fungi presented in soil on bast stems. Unfortunately, it is a time-consuming process with inconsistent outcomes, in which fibre quality depends on geographical conditions; even though this method is found cheaper, it yields higher fibre content and creates less pollution.

On the contrary, chemical retting produces a more controllable bast fibre quality within a short retting duration. High processing cost, unfavourable colour and deteriorated tensile strength of the retted bast fibre have forced to discover a better retting method. Enzyme retting has been so popular due to its mild process conditions, specificity and high selectivity with no chemical present. The characterisation of enzyme-retted bast fibres shows comparable quality as water-retted fibres. However, some studies found lower fibre strength because of continued activity of cellulase in the enzyme mixtures. Therefore, controlling retting duration is essential to avoid over-retting.

i) Water Retting

Water retting is the oldest historical retting method. This process was once popular in producing quality-retted bast fibres. The most widely practiced method of retting is water retting which is performed by submerging bundles of stalks in water. The water, penetrating to the central stalk portion, swells the inner cells, bursting the outermost layer, thus increasing absorption of both moisture and decay producing bacteria. Retting time must be carefully judged; under-retting makes separation difficult, and over-retting weakens the fibre. In double retting, a gentle process producing excellent fibre, the stalks are removed from the water before retting is completed, dried for several months, and then retted again. Natural water retting employs stagnant or slow-moving waters such as ponds, bogs and slow streams and rivers. The stalk bundles are weighted down, usually with stones or wood, for about 8 to 14 days depending upon water temperature and mineral content (Fig. 2.3).

Fig. 2.3 Water retting

ii) Dew Retting

During the dew retting process, stems of the plant are being cut and evenly distributed in the fields, where the presence of bacteria, sunlight, atmospheric air and dew causes breakdown of stem cellular tissues and adhesive substances that surrounded the fibres. The places that have a warm day and heavy night dew are preferred for the dew retting process to promote colonization of fungi. Fungi colonization is considered as the major enzymatic active method responsible for dew retting. However, recent studies have shown interest on the interaction of the bacterial and fungi community during dew retting. This is a common method in areas with limited water resources. It is most effective in climate with heavy night time dews and warm daytime temperatures. The harvested plant stalks are spread evenly in grassy fields, where the combined action of bacteria, sun, air and dew produces fermentation, dissolving much of the stem material surrounding the fibre bundles. Within two to three weeks, depending upon climatic conditions, the fibre can be separated. Dew-retted fibre is generally darker in colour and of poorer quality than water-retted fibre.

iii) Chemical Retting

Chemical retting degrades non-cellulose components, but cellulose degradation happens when over-retting. Retting duration, chemical concentration and retting temperature have been reported that affect the quality of retted fibres. The uses of chemicals and production of wastewater have increased the cost as well as polluting environments. Using reused wastewater for continued chemical retting process, an addition of 50% of chemicals is needed to obtain comparable physical and mechanical properties of the retted fibre as the first retted batch. This method has essentially reduced the cost as well as wastewater generation. Effective chemical retting reduces retting time to 2 h. The chemical-retted fibre showed positivc results towards mechanical properties with reducing flexural rigidity, linear density and diameter. Softer fibres are obtained by removing impurities. However, contaminated water showed exorbitant chemical oxygen demand, biological oxygen demand, total dissolved solids, sulphide content and blackish brown colour. Therefore, a new treatment viz., electrocoagulation has been introduced to reduce the contamination of wastewate and the outcome was satisfied. It provides quicker, simpler and economical treatment by using electrocoagulation. Duration of 90 min gives noticeable reduction in contamination of compounds as well as it is cheaper in cost.

iv) Enzyme Retting

Enzyme retting process has been introduced for some years back as a potential substitution to the above retting methods. Dew retting process is often constrained by poor and inconsistent fibre quality as well as geographical region, which requires optimum temperature and moisture to promote microbial growth. Therefore, it is less efficient in countries with dry climate. On the contrary, enzyme retting showed promising results in Europe when commercial pectinase-rich enzyme is used in the retting of the flax fibre. The retted fibre has higher yield and comparable quality to the water-retted fibre. The retted stalks called straw are dried in open air or by mechanical means and are frequently stored for a short period to allow "curing" to occur, facilitating fibre removal. Final separation of the fibre is accomplished by a breaking process in which the brittle woody portion of the straw is broken, either by hand or by passing through rollers, followed by the scutching operation, which removes the broken woody pieces by beating or scraping. Some machines combine breaking and scutching operations. Waste material from the first scutching, consisting of shives and short fibres, is usually treated second time. The short fibre obtained is frequently used in paper manufacture and the shives may serve as fuel to heat the retting water or may be made into wallboard and to make rope.

FLOSSES

Many trees and shrubs in India produce silky flosses within their fruit. Flosses are generally used for stuffing mattresses, pillows, cushions, upholstery and packing. They are also used for the manufacture of life belts. Some of the important floss yielding species found in the forests of India are as follows.

***Ceiba pentandra* (Kapok tree)** is a medium sized deciduous tree found widely distributed in the hotter parts of western and southern India and Andaman Islands. The tree is commonly called as silk cotton or kapok tree. The floss is obtained from the pods of the plant. Unopened pods are collected from the tree with hand and are allowed to dry in the sun. After drying, on beating with mallet, the pods are split open. The floss is taken out along with the seeds and dried in sun. On beating lightly, the seeds yield white or pale yellow colour floss. The floss is light brittle, elastic and lustrous. The yield of the floss varies with the age of the plant. As the age advances, the yield per tree also increases. It finds use in bedding and upholstery industries. It is used in stuffing pillows, cushions and mattresses. It is also used for making padded surgical dressing. Due to water qualities, it is used for manufacturing life jackets and belts, waist coats, fishing and aviation suits and other naval life saving devices. Recently, with the help of modern machines, the floss of this species has been spun either alone or in admixture with cotton into yarn. This yarn can be dyed or bleached like cotton. It is used for insulation of refrigerators, cold storage plants, offices, theatres, motor cars and aeroplane for sound insulation. It is used in aeroplane cabins, studios, hospitals, auditorium and theatres (Fig. 2.4).

Fig. 2.4 Floss from *Ceiba pentandra*

Himalayan Poplar (*Populus ciliata*) is a large deciduous tree with a straight bole growing extensively in the Himalayan region between 1500-2600m. It grows on early successional condition especially along watercourses in the Central Himalaya. The ratio of male and female tree varies from site to site. The new leaves appear towards the end of March or early April. The catkins appear immediately before the young leaves. The female trees have new flush of leaves simultaneous with the flowering while male trees do not unfold leaves until the dehiscence of pollen is over. The seed dispersal takes place from the end of May to June depending upon the climate of the locality. The species yields abundant floss which is enclosed around the seeds in the pods. The floss is used for filling pillows and mattresses.

Semal (*Bombax ceiba*) is medium to large sized deciduous tree, typical of alluvial savannah type of forests. The species is found growing sporadically in mixed deciduous forest of sub-Himalayan region and moist deciduous forests of west coast. Fibres are obtained from the inner side of the capsule. It is slightly brownish yellow in colour. The floss from this tree is known as Indian kapok. The green capsule of this tree yields soft and strong floss about 4.5-6.0 kg per tree in full bearing. Kapok is similar to the true Java kapok and is used in life saving appliances and other general flotation purposes. The floss is used for stuffing quilts, mattresses, cushions and pillows. Articles stuffed with kapok are said to be vermin proof. It is also used as an insulating material for refrigerators, sound proof covers and walls. It is considered to be the best for making padded surgical dressings. In the recent years, with the help of the modern modified carding machine, the floss can now be easily spun in to yarn.

Akund floss (*Calotropis gigantea*) is a common shrub found in the Sub-Himalayan regions, Central part of India and Southern zone. The seeds bear fine, soft, glossy and resilient floss. Through inferior to Indian kapok in strength and resilient properties, it has good buoyancy. Floss from this species is used in stuffing mattresses, pillows, etc. It can be spunned either after chemical treatment or mixed with cotton but it cannot be bleached.

Yellow cotton tree (*Cochlospermum religiosum*) is a small or medium sized deciduous tree, commonly called as the yellow cotton tree. It is common in hot dry and stony regions. It is found almost throughout India. The seeds of this species are covered with soft and resilient floss which can be used as substitute for kapok. It is mostly used for stuffing mattresses, pillows, cushions and life belts.

Suggested Readings

Ibrahim H Mondal 2021 *Fundamentals of Natural Fibres and Textiles,* New Delhi: Woodhead Publishing Pvt. Ltd

Ryszard M Kozlowski and Maria Mackiewicz-Talarczyk 2020 *Handbook of Natural Fibres: Volume 1: Types, Properties and Factors Affecting Breeding and Cultivation,* New Delhi: Woodhead Publishing Pvt. Ltd

Tribhuwan Mehta 2008 *A Handbook of Forest Utilization*, Delhi: Periodical Expert Book Agency

Uma Shankar R Hiremath Ankila J Hiremath Gladwin C Joseph and Rai D Nitin 2009 *Non-timber forest products: conservation, management and policy in the Tropics,* Bangalore: GKVK Campus

CHAPTER 3
Gums and Resins

M.P. Divya and C.N. Hari Prasath
Tamil Nadu Agricultural University, Coimbatore

Gums and resins occupy a prime place among Non-Timber Forest Products (NTFP) and are known to mankind since time immemorial. Gum trees are economically important and found in tropical moist and dry deciduous forests. Mostly gums and resins are plant exudations, partly as a result of injury to the bark or stem. Generally, gums are exudates by the stem. Only a few gums are obtained from roots, leaves and other parts of the plant. Gums are supposed to be the result of the disintegration of internal tissues in which cellulose is decomposed into more or less a viscous substance through a process called gummosis, which exudates from cracks and wounds of the stem. Gums contain a large quantity of sugar and are closely related to pectin. Gums are generally soluble in water and soften in cold water. They are insoluble in alcohol and other organic solvents. They decompose completely on heating without melting and tend to become charred. Mostly gums exude from plants in liquid form and on exposure, dry up into translucent, amorphous and ovoid tear shape bodies which stuck to the bark of the stem or branch.

GUMS

Tree gums comprise of polysaccharides or their derivatives, with the formation of true gums being due to the disintegration of internal plant tissues. This takes place primarily through the process of gummosis brought about by the decomposition of cellulose. Gums are used for variety of purposes. Gums which are colorless and possess greater adhesive power and more viscous in nature are considered to be the most valuable among gums. These gums are generally utilized for classification of liqueurs, finishing of silk and preparation of quality water colours. Intermediate grade gums are used in confectionary, pharmaceutical, printing inks, sizing and finishing of textile fabrics and dyeing. The low grade gums are used as adhesives, in calico printing, sizing of paper and also find use in paint industry. In pharmaceutical and cosmetic industries, gums are used as demulcent or emollient or emulsion to bind mixtures in creams, lotions and ointments. Gums are used in nutritious food (Table 3.1).

Table 3.1. Main Characteristics of Gums and Resins

Gums	Resins
Gums are either hydrophobic or hydrophilic	Resins are hydrophobic and they have a characteristic lustre, are ordinarily fusible and when ignited, resins burn with a smoky flame
Dissolve or soften in cold water	Soluble in ordinary solvents like alcohol, ether and turpentine
Insoluble in alcohol or other inorganic solvents. They are also insoluble in oils or organic solvents such as hydrocarbons, ether and alcohol	Insoluble in water
The mixtures are often complex and on hydrolysis yield simple sugars such as arabinose, galactose, mannose and glucuronic acid	Non-nitrogenous compounds
On exposure to air, gums dry up into translucent, amorphous, tear-shaped bodies or flakes	They are brittle, amorphous and transparent or semi-transparent

Kinds of Gums

True gums are generally divided into two main classes or groups viz., i) Soluble Gums. Its characteristic properties are soluble in water, more / less transparent, viscous in nature and adhesive solution, eg. Indian Gum Arabic obtained from *Acacia nilotica* and ii) Insoluble Gums. Its characteristic properties are that it absorbs water and swells into thick jelly with sufficient water, breaks down into very thick translucent solution eg: Gum Karaya obtained from *Sterculia urens*

Gum Yielding Species

Several species belonging to genus *Acacia, Anogeissus, Mimosa, Albizia*, etc. yield gums of importance. Some of the important gum yielding trees are given below.

Gum Acacia (*Acacia nilotica*) is found in many dry regions of the country and it is the main source of the commercial Indian gum Arabic or babul gum (*Acacia nilotica*). Exudation occurs spontaneously during the dry season when they harden as round or oval tears. It is composed of galactoarabinan which gives on hydrolysis L-arabinose, D-galactose, L-rhamnose, D-glucuronic acid and 4-Omethyl-Dglucronic acid. The gum which initially varies in colour from very pale yellowish brown to dark reddish brown depending on the quantity of tannins in the exudates.

The gum obtained from this tree is lighter and very viscous. This gum has been used in textile, cosmetic, ink, mucilage, paste, polish medicines and confectionery industries and as a glaze in painting.

Gum Arabic (*Acacia senegal*) is a dried gummy exudation of *Acacia senegal.* The gum dries up on the tree and is removed after about a month. The gum may be used either separately or mixed with other acacia gums. It is a mixture of calcium, magnesium and potassium salts of arabic acid. Gum Arabic is composed of galactose, arobinopyranose, arabinofuranose, rhamnose, glucuronic acid and 4-0 methyl glucuronic. It is highly soluble in water and maximum viscosity obtained at neutral pH. Gum arabic is a complex mixture of glycoproteins and polysaccharides, predominantly polymers of arabinose and galactose. It is soluble in water, edible and used primarily in the food industry and soft-drink industry as a stabilizer. Gum arabic is a key ingredient in traditional lithography and is used in printing, paint production, glue, cosmetics and various industrial applications including viscosity control in inks and in textile industries. It is used in the food and pharmaceutical industries as a stabilizer, thickener, suspending and binding agent. It is also used as a binding agent in the manufacture of confectioneries, dairy products, beverages, cottonseed oil emulsion and tablets.

Gum Karaya (*Sterculia urens*) exudation obtained from the incised trunk of the stems and branches of *Sterculia* species mostly from *Sterculia urens*, a large tree in central India. This tree is found in the tropical deciduous forests of central and south India. Gum karaya is the trade name of the gum obtained from this tree, it is not soluble in water. It traditionally tapped by cutting or peeling back the bark or by making deep gashes at the base of the trunk with an axe. Exudation takes place from the whitish or pinkish brown bark on injury or during tapping which is done by making two blazes on its trunk. It consists of high molecular weight acetylated polysaccharides which on hydrolysis yields galactose, rhamnose and arabinose. It contains 0.17% nitrogen content in amino acid form. It possesses high viscosity and low solubility in water. It can absorb water and swell to 60-100 times its original volume, a property useful in bulking agents. The morphological characteristic of the gum karaya exudates shall be white to the amber colour in the form of tears of variable size or in broken irregular pieces. It shall possess a slightly acetous odour and a mucilaginous and slightly acetous taste. It is one of the least soluble gums used widely in the petroleum and gas industry, textile industry, paper industry, pharmaceutical industry and also used in several other products due to its stability, high viscosity and suspension properties. The gum karaya is well suited for stabilizing low pH emulsions in sauces and dressings. It is also used to adulterate Gum tragacanth due to their similar physical characteristics. As a food additive it has E number E416.

Gum Albizia is derived from trees of the genus *Albizia* which is formed as round elongated tears and colour ranging from yellow to dark brown. It contains L-arabinose, D- galactose, D-glucuronic acid, D- mannose and 4-0-methyl glucuronic acid and L-rhamnose with slight acetylation and methylation. It is a complex salt of calcium, potassium, magnesium and sodium in decreasing proportions. It is soluble in water, forms as colourless mucilage with a bland taste. Viscosity of *Albizia* solution is greater than that of gum Arabic solution but is less than the viscosity of gum karaya. The promising substitute for the gum Arabic is used as a natural emulsifier in the pharmaceutical and manufacturing industry. It is used as coating material, binding agent in tablet formulation and also in food industry.

Gum Tragacanth (Genus *Astragalus*) is a natural gum obtained from the dried sap of several species of middle eastern legumes of the genus *Astragalus*, including *A.adscendens*, *A.gummifer*, *A.brachycalyx* and *A.tragacantha*. Gum tragacanth is a viscous, odorless, tasteless, water-soluble mixture of polysaccharides obtained from sap that is drained from the root of the plant and dried. The gum seeps from the plant in twisted ribbons or flakes that can be powdered. It absorbs water to become a gel, which can be stirred into a paste. The major fractions are known as tragacanthin, highly water-soluble as a mucilaginous colloid and the chemically related bassorin, which is far less soluble but swells in water to form a gel. It is a complex mixture of acid polysaccharides containing galacturonic acid, galactose, fucose, arabinose, xylose and small amount of rhamnose and glucose. It swells in water to give a thick gel which shows typical pseudo plastic behavior. It is used extensively in food products with E number E413. Gum tragacanth is also used in incense making as a binder to hold all the powdered herbs together.

Gum Ghatti (*Anogeissus latifolia*) gum oozes out naturally from the bark through injuries and wounds mostly in summer and is collected manually. In some places, artificial incisions are made in the tree bark to increase the gum yield. These incisions are made carefully so as not to permanently injure or kill the tree. Tree yields a valuable gum called gum ghatti. An improved tapping method based on application of ethephon yield more gum. Maximum quantity of the gum is collected during the summer months i.e., from March to mid of June. During this time, as the weather gets warmer, the yield increases. Normally the largest crop is picked in April. Ghatti is an amorphous, translucent, water-soluble gum. The colour of the gum varies from whitish-yellow to amber; though the presence of impurities sometimes imparts a brownish color to the gum. Gum ghatti is partly soluble in water and forms colourless mucilage. Three grades of gum ghatti viz., No.1, No.2 and Unassorted grades are available which are exported to various countries. The No. 1 grade is light in colour with low levels of ash and high viscosity. No. 2 grade is light amber to brown and the Unassorted grade is dark brown to nearly black in color. A BIS specification (IS 7239:1974) exists for food grade gum ghatti. It occurs naturally as a calcium and magnesium salt of a complex polysaccharide's acid- complex. The main chemical constituents of Gum ghatti consist of L-arabinose, D-galactose, D-mannose, and D-glucuronic acid residues. On dispersion in water, gum ghatti forms viscous solutions, intermediate between those of Gum arabic and Gum karaya. Dispersions have emulsifying and adhesive properties. It is used as a food additive besides being used in medicines, textiles and adhesives. Gum ghatti is used as an emulsifier and stabilizer in beverages and butter containing table syrups.

Neem gum is a natural gum extracted from neem tree (*Azadirachta indica*) by induced or natural injury. It is generally not tapped because of the meager amount of exudation. Direction of ethephon and paraquat (1,1'-dimethyle 4,4' bipyridium salt) into the sapwood of neem mediated copious gum exudation. The treatment induce the formation of gum ducts and cavities in the sapwood. Neem gum is clear, bright and amber-coloured material non-bitter in taste and is soluble in cold water. Chemically, it contains mannose, glucosamine, arabinose, galactose, fructose, xylose and glucose. The gums are polysaccharides, composed of polymers of sugars and oxidized sugar linked together by glycosidic bonds. It is used as a bulking agent and for the preparation of special purpose food. Pharmaceutical applications of neem gum is as tablet binder, thickening agent, slow-release agent and film coating. There is also some use for enhancing solubility. Neem gum is also used in silk dying.

Tamarind Seed Gum is obtained from the endosperm of the seed of *Tamarindus indica*. Tamarind gum/Tamarind kernel powder came into commercial production in 1943 as a replacement for starch in cotton sizing in Indian textile market. Tamarind gum polysaccharide, the purified product is a free flowing, pale, creamy white powder without taste or odour. Highly branched carbohydrate polymer. Branching from Polyglucopyranosyl of one or two D- Xylopyranosyl units capped by D-galactosyl, D- Xylopyranosyl and L-arabinosyl. Tamarind gum possesses hydrophilic, gel-forming and bio-muco adhesive characteristics and the aqueous solution of tamarind gum generates a stiffer conformation with the large volume of occupancy quickly hydrates in cold water. It reaches the maximum viscosity after boiling for 20-30 min. Tamarind gum has been applied in the field of drug delivery systems due to its unique characteristics such as high viscosity, broad pH tolerance, non-carcinogenicity, muco adhesivity, biocompatibility and high drug holding capacity. Due to its unique physico-chemical characteristics, the uses of tamarind gum as a suspending agent, emulsifier, thickener, gelling agent, film former, binder, disintegrating agent, drug-release modifier, matrix former, muco-adhesive agent, etc.

Moringa gum (*Moringa oleifera*) is one of the newly recognized gum. Moringa gum is the exudate from the bark of *Moringa oleifera.* The chemical constituents of the gum exudate contain L-arabinose, D-galactose, D-glucuronic acid, L-rhamnose, D-mannose and D- xylose. *Moringa oleifera* is a small genus of quick growing tree distributed in India. The bark of moringa tree is thick, soft, fissured and corky, becoming rough. When a moringa tree is injured, its stem exudes a moringa gum with a bland taste, which is initially whitish-grey in colour but changes to reddish-brown to brownish black on the exposure to the atmosphere. The pharmaceutical and other industrial applications of the gum are gauged as stabilizers, binders and muco-adhesive. Globally it has been esteemed as nutraceuticals, therapeutic and cost-effective properties.

Miscellaneous gum yielding species

Albizia chinensis, Albizia odoratissima, Albizia procera, Anacardium occidentale, Bauhinia purpurea, Bauhinia racemosa, B. variegata, Bursera bipinnata, Caesalpinia spinosa, Ceratonia siliqua, Chloroxylon swietenia, Cochlospermum religiosum, Grewia mollis, Limonia acidissima, Mangifera indica, Prosopis juliflora, Terminalia alata and *Terminalia bellirica.*

GUM TAPPING METHODS

i) Selection of trees

Trees of and above 90 cm. in girth at breast height are tapped. Trees of lower girth classes are not tapped since any damage to them may adversely affect their growth. Middle aged trees yield more gum. Selected trees are marked and numbered. Tapping, which is the practice of wounding/incision of trees to facilitate the flow of exudes from the trees. Tapping and gum collection takes place during the dry season. Wounds are healed when vegetative growth is initiated during the rainy season. Tapping directly affects both quality and quantity of the products. Wounded either by cutting or piercing by local hand tools such as scissors, peelers, knives, axes, sickles, stones, metal or any sharpened material. Tapping is done twice a year although a tree might be tapped up to five times a year depending on localities and tree species. The gum exudes from the trees all the year round. The flow is more in hot weather. Tapping may be done from October to June. No collections should be made during rainy season as most of the exuded gum is either washed away or become darker in color. The factors that affect the gum yield are i) stage of harvesting (maturation, flowering stage), ii) time of harvesting (early morning/ evening), iii) rate of drying (to avoid decomposition) and iv) drying temperature (to avoid decomposition), v) moisture content after drying (to avoid molds growth), vi) storage conditions (to prevent hydrolysis, oxidation, infestation) and vii) storage time before processing (loss of oil) (Fig. 3.1).

Fig. 3.1 Gum exudation

ii) Tapping methods

a) Blazing

Tapping method varies from region to region. Making deep and very large blazes of about 6 cm. or so results in girdling of trees. Blazing /stripping off the tree bark or using gum inducer method are employed. The initial blaze of 15 x 10 x 2.5 cm to 3 cm deep (6 "long 4" wide and one to 1.2 inch) may be made on trees with sharp knife having 7.5 cm (3") wide blades. The blaze should be just deep enough to take out 0.5 cm. (0.2") deep wood portion below the cambium. Deeper blazing may damage the tree since it will delay the process of healing. In the subsequent years, the same blazes are tapped vertically and laterally without giving any chance for the tree to heal up its wounds.

b) Specialized tools for gum tapping

Acacia senegal and ***Acacia nilotica***: Tapping methods have been developed which do not damage the trees, and handling and cleaning practices have been optimized to produce a superior quality product. The older methods of making small incisions into the tree with an axe have largely been replaced by one which utilizes a specially designed tool, "Sonki". This has a metal head fixed to a long wooden handle.

Acacia polyacantha: The highest gum production is obtained when the trees are tapped by using Makmak tool. Tribal uses a special tool called "Saluli" for peeling off the stem bark and for collection of salai gum, a special tool called "Gaantri" and for the tapping gum-butea "Bill hook" is used.

iii) Grading

Processing and grading of raw gum are done manually or mechanically and it is almost the same in principle. In various gum grading centers in Madhya Pradesh, raw gum is cut with choppers into 2.5 to 5 cm pieces. These are dried in sun and graded further by breaking and sorting according to size, colour and impurities. Drying takes about 5 to 15 days depending on the initial moisture of the raw gum and also weather conditions. As per Indian standard institution specification (I.S. 5025-1969), gums are classified into three grades i.e. Grade I-white to amber color, Grade II- reddish to pale yellow and Grade III -brown to black. The gum passing upto 4 mesh size sieve is A grade and 5-16 is B grade. Inspection of gum grades particularly A and B meant for food and pharmaceuticals is done on aluminum plate to detect rodent excreta and hairs at the time of final grading. Grading is done using electrically operated disintegrating plant with arrangements for grading gums into different sizes and colours. Such plants crush about 5 to 6 quintals of raw gum per hour. Another machine which has a decorticator with fans for wind blowing is used for separating bark adhering to gum (Fig. 3.2).

Fig. 3.2 Gum grading

iv) Storage

The raw gum is generally stored in old gunny bags (capacity 70 -75 kg). The processed gum is stored and marketed in first quality gunny bags (80 kg). Gum should be stored under cool and dry conditions. Gum in powdered form is to be stored in well closed containers to prevent absorption of moisture. Grades and ungraded gum must be stored separately in clean hygienic premises and prevent them from being mixed with rodent excreta and bird dropping. Depots and godowns should be rat-proof and rain-proof. Since, it is an edible product, sanitary conditions in storage must be ensured and for this purpose it may be necessary to prescribe inspection of export consignments. Gum bags as they come from center of production should be checked as per samples of each grade maintained under proper supervision and each packet should be labeled with the grade it contains. Gum should be preferably packed in waterproof containers and the containers sprayed with insect and rodent repellent.

v) Gum yield

The gum yield of important gum yielding trees is given in Table 3.2.

Table 3.2 Gum yield of different tree species

Gum yielding species	Tapping methods	Season	Gum yield (kg / tree)
Gum Karaya	Blazing method	October to June	2.5 to 5
Gum Arabic	Bore hole method + chemical method	March to June	0.4 to 7
Babul gum	Bore hole method + chemical method	March to May	0.4 to 1.1
Jhingan gum	Bore hole method + chemical method	April to June	2 to 4
Gum Ghatti	Bore hole method + chemical method	April to May	1 to 2
Salai gum	Square shape tapping + chemical method	March to May	1 to 2.5
Guggul gum	Bore hole method + chemical method	November to February	0.04 to 0.32
Neem gum	Bore hole method + chemical method	April to June	1 to 2
Phulai gum	Bore hole method + chemical method	March to June	0.1 to 0.9

Gum yielding species	Tapping methods	Season	Gum yield (kg / tree)
Cashew nut gum	Bore hole method + chemical method	March to May	0.03 to 1.2
Semla gum	Bore hole method + chemical method	April to June	1.5 to 2
Siris gum	Bore hole method + chemical method	March to June	0.4 to 0.7
Khair gum	Bore hole method + chemical method	April to June	1 to 2

Applications/ Uses of tree gum

- Gum is used as a suspending and emulsifying agent
- In cosmetics, it imparts spreading, viscosity and protective characteristics to lotions and protective cream by stabilizing the emulsion.
- Gums are used in the manufacture of adhesives for postage, stamps, paints and inks and have been applied in lithography and textile industry as a sizing and finishing agent
- Used in pharmaceutical preparations, confectionery, sweet meats and cosmetics
- Used as an emulsifier and as a stabilizer in frozen dairy products, etc.

RESINS

"Resin is a solid or highly viscous substance of plant or synthetic origin that is typically convertible into polymers". Resin secretion occurs in special cavities in many plant species. They are formed in the specialized structures called passages ducts. Resins exude or ooze out from the bark of the trees and tend to harden on exposure to air. Natural resins of particular importance to the furniture coatings are rosin, damar, copal, sandarac, amber and manila. A sticky flammable organic substance, insoluble in water, exuded by some trees and other plants typically convertible into polymers. Resins are usually mixtures of organic compounds. Plants secrete resins for their protective benefits in response to injury. The resin protects the plant from insects and pathogens. Resins are formed as oxidation products of various essential oils and are very complex and varied in chemical composition. It frequently oozes out through the bark and hardens on exposure to air. Tapping is necessary in order to obtain sufficient amount of resin for commercial value. Resinous substances may occur alone or in combination with essential oils or gums. Resin production is widespread in nature, but only a few families are of commercial importance. Resins probably save the plant by preventing decay, which is due to their high antiseptic qualities. They may also lower the amount of water loss from the plant tissues.

The resin, which comes out is collected. Most plant resins are composed of terpenes. Specific components are alpha-pinene, beta-pinene, delta-3 carene and sabinene, the monocyclic terpenes, limonene and terpinolene and smaller amount of tricyclic sesquiterpenes, longifolene, caryophyllene and delta-cadinene. Some resins also contain a high proportion of resin acids. Rosins on the other hand are less volatile and consist of diterpenes among other compounds. The characteristics of resins are i) insoluble in water, ii) soluble in ordinary solvents like alcohol, ether and turpentine, iii) brittle, amorphous and are transparent or semi-transparent and iv) It has a characteristics luster, are ordinarily fusible and when ignited, resins burn with a smoky flame.

Classification of Resins

The resins are classified into three types i) Hard Resins, ii) Oleo Resins and iii) Gum Resins.

i) Hard Resins

- Contain no or little essential oil
- Solid, transparent and have no specific taste and smell
- These are the best source of varnishes
- Dammar-used for varnishes (obtained from trees of families Burseraceae and Dipterocarpaceae)
- Lacquer-natural varnish quickly hardens, obtained from *Melanorrhoea usitata* and *Rhus verniciflua*
- Mastic - very expensive used for coating metals and pictures
- Sandarac- soft resin obtained from *Colitis quadrivalvis* is used for coating labels, leather, metals and negatives.

ii) Oleo Resins

- Contain considerable quantity of essential oils together with resinous material. The most common compounds of oleoresin are α-pinene @ 19.7 %, β-pinene @ 13.3 %, γ-terpinene @ 10.2 % and β-caryophyllen @ 7.8 %
- It is in liquid form
- It has distinct flavour or aroma
- Turpentine is obtained from coniferous trees

- Balsam-contain benzoic or cinnamic acid, hence highly aromatic. Distillation yield essential oils. It is used in medicines and pharmaceuticals and perfumeries.
- Copaiba from genus Copaifera. It is used as fixative in scenting soaps and perfumes
- Elemi- soft resins from genus *Protium* and *Canarium.* It is used for hardening or toughening varnishes.

iii) Gum Resins

- Mixture of both gums and resins. Hence it has a character of both the group.
- It also contain small quantities of essential oils.
- It is obtained from plants grown in dry and arid regions.
- Asafoetida-strong aromatic & bitter taste. Exudes from roots of *Ferula asafoetida*, when the stem is cut off. It adds flavour to curries, sauces and used as medicine.
- Frankincense or Olibanum from *Boswellia* spp. It is used as incense in churches and homes.
- Galbanum is obtained from stem of *Ferula galbaniflua,* used in medicines.
- Gamboge is a hard, brittle, yellow coloured resin from *Garcinia* spp. It is used as pigment for water colours and medicines.
- Myrrh. It is a most valuable gum resins obtained from *Commiphora* spp. It is used for preparation of incense and perfumes and for embalming.

Resin yielding species

Black dammar (*Canarium strictum*) is a dioecious tree distributed across parts of India, Myanmar and China. Flowers are insect pollinated and trees can grow upto 40 m tall and are found in moist deciduous to semi-evergreen forests at an altitude ranging from 750 m to 1400 m. *Canarium strictum* yields the resin known as 'Black dammar' and the local names in India include mand dhuup, raal dhuup, karunkungiliyam and sambrani. The resin from *Canarium strictum* is not secreted from the bark itself; the tree needs to be incised for the secretion of resin. Tapping is done by making vertical incisions on the bark in a belt about 1.8m from the bottom and then lighting a fire around the base of the tree; the bark and outer layers of the wood are thus damaged. The resin flows out after two years, and this flow lasts for about six months every year from November to April and continues for about 10 years.

The resin that oozes out is viscous and hardens into a translucent and stalactite mass. When it reaches to the market, it is dark black, brittle with nodular lumps. It can be easily ground into a light, dust coloured powder. It is partially soluble in alcohol, acetone and completely soluble in benzene and turpentine. Traditionally the resin is used to treat rheumatism, asthma, venereal disease and for chronic cutaneous diseases such as psoriasis and pityriasis and as a liniment in rheumatic affected areas. On the other hand, topical application of stem bark powder of *Canarium strictum* is reported as a mosquito repellent. Apart from medicinal uses, resins of *Canarium strictum* are commercially harvested for their economic importance to local communities, used as incense and for the manufacture of varnish, lacquers, and paints. It is chiefly used in the manufacturing of varnishes, caulking boats and bottling wax and as a substitute for burgundy pitch in plasters. The annual yield of this resin is about 80 tons per annum.

White dammar (*Vateria indica*) is a large evergreen tree, indigenous to the evergreen forests of the Western Ghats from North Kanara to Kerala, belongs to Dipterocarpaceae family. It is a multipurpose tree which has economic and medicinal importance. Resin from *Vateria indica* is known as 'White dammar'. Semi-circular incisions on the stem through the cork cambium are made, up to the surface of the sapwood and resin obtained. Blazes or cuts are so spaced as to cause the least damage to the tree. The resin starts oozing from the incisions in 3 to 4 days and continues for 60 to 90 days. Comparatively, a high yield of resin can be obtained if the bark is scorched by lightning fire around the base of the tree, but this method is generally not practiced to prevent the fire. The dried portions of the cuts and blazes require fresh cuts for more yields. The resin, when soft is called *Piney varnish* but when hard is called *Indian dammar*. The piney resin is sold in the market in three forms viz., compact form, cellular form and dark coloured form. The compact form is solid lumps and is regarded as the best quality and it is very hard and bright orange to dull yellow. The cellular form is full of air bubbles and gives the appearance of a cellular structure and is light green to yellow or white and has shining appearance. The third form occurs in cavities of old and moribund trees or dead trees and is of inferior quality. It is used in Indian medicine as a tonic for throat troubles, chronic bronchitis, piles, diarrhoea, rheumatism, tubercular glands, carminative and expectorant. Apart from medicinal uses, it has long been used as incense and for making varnishes.

Rock dammar (*Hopea odorata*) is a medium-sized to large evergreen tree naturally found in the Andaman Islands in small groups and cultivated in West Bengal. This tree yields commercial resin called as 'Rock dammar' by natural exudation but it is not usually tapped for commercial purpose. The tree exudes as large, yellow, irregular tears which harden on exposure to the atmosphere. It is fairly hard and completely soluble in turpentine. Harvesting of the resin commences when the bole is around 25 cm in diameter (approx. 20 years old). Triangular cuts (becoming circular with age) are arranged in vertical rows around the trunk. The cuts are several centimeters wide at first, but become enlarged at every tapping and eventually become holes of 15 - 20 cm in depth and width. The average number of holes for a tree about 30 m tall and 60 -80 cm in diameter is 9 - 11 in each of 4-5 vertical rows. For the higher holes, the tapper climbs the tree supported by a rattan belt and uses the lower holes as footholds. The exuded resin is allowed to dry on the tree before it is collected. The frequency at which refreshening the cut varies from once in a week to once in a month depending on the diameter and health of the tree and the tapping can continue for 30 years. Compare to other dammars, rock dammar is considered a second quality dammar and is used in the preparation of varnishes, painting pictures, caulking boats and other medicinal purposes. It is an ingredient of inks, lacquers, oil paints and varnishes and is used as a glazing agent in foods.

Sal (*Shorea robusta*) is a large deciduous tree, found in the forests of central, northern and eastern India. It yields resin commonly called as the Sal Dammar / Lal Dhuma. The method of tapping employed is to cut 3 to 5 narrow strips of bark, 90 to 120 cm above the ground. In about 12 days, the grooves become filled with the resin, which oozes out a whitish liquid that becomes brown soon on drying. The cavities of the grooves freshened, after which the exudation continues and the resin collected. The process is repeated several times during the year. The maximum yield of sal dammar is obtained from September to November. An approximate annual yield of resin is about 5 kg in a mature tree. This tree naturally exudes the resin if that tree is naturally injured by a heartwood borer. The indigenous tapping system causes permanent damage to sal tree. So in reserve forest, the tapping has been prohibited because sal wood is commercially important timber species. The upper blazes yield more resin and the maximum yield is obtained from September to November and fortnightly freshening yields more resin than monthly freshening. The sal dammar occurs in a nearly cloudy, brittle, pale yellow and has aromatic odor. The resin is used for the preparation of varnishes, incense making, disinfectant fumigant, inferior quality paints, skin ointments and for caulking boats.

Salai (*Boswellia serrata*) is a large sized deciduous tree majorly found in dry deciduous forest areas of throughout India. *Boswellia serrata* on naturally or tapping, exudes gum-oleo-resins which is commercially known as Salai gum or Guggul or Indian olibanum. The boswellic acids, the active compounds in *Boswellia serrata* gum resin extract, are potent anti-inflammatory agent and are specific non-redox inhibitors of 5-Lipoxygenase (5-LOX). The trees of 90 cm girth middle aged, sound and vigorous trees give a comparatively good yield. The bark shaved off from the trunk in trees at 0.75 m above ground height is suitable for resin tapping. The phloem which contains resin canal and ducts are exposed for exudation of resin. The fresh exudation from the punctured resin-ducts comes in 5 to 8 cm long tears, resembling Canada balsam in colour and consistency. The first collection of resin is made after 7 to 15 days of tapping. The resin hardens in four days and it is golden in colour and transparent in nature. Tapping is done during the end of winter to the summer sometimes extends from November to June or July. The blaze is made after every fourth and fifth day to facilitate resin exudation. The number of blazes required depends on the girth of the trees and initially for a tree of 90 cm, one blaze may be made and every increase of 50 cm girth, one more blaze may be added. In *Boswellia serrata*, the tapping does not create any damage to wood. The borer attacked or hollow trees also exude a good amount of Salai Guggul. The yield has generally been found to be very poor in moist conditions and from trees that are old with black bark, suppressed and dwarfed. A well grown tree yields about 2 to 2.5 kg of gum-oleoresin in a year. This oleoresin is used as an incense, fumigant, medicinal purposes for the treatment of rheumatism, nervous diseases as a diaphoretic, astringent and skin ointments.

Gurjan (*Dipterocarpus* spp.) is a large sized evergreen tree naturally found in moist tropical forests of Andaman Islands and some part of Assam. Gurjan is a source of the Gurjan oil in West Bengal and commercial kanyin oil in Myanmar. Above 1m from the ground area is selected for tapping. A cone shaped cavity is made on the trunk of the tree and fire lighted to char the cut surface. A tree of 1.8 m girth yields about 10 kg of resin in one season i.e., November to May. This oleoresin is used in the preparation of varnishes and lithographic inks, providing an anticorrosive coating to iron for the preservation of timber and bamboo and for caulking boats. The other Dipterocarpus species like *Dipterocarpus alatus* and *Dipterocarpus macrocarpus* also produce the minor quality oleoresins.

Pine (*Pinus* spp.) is a large evergreen and endangered species, commonly found in the evergreen forest of Western Ghats from South Kanara to Kerala. The tree yields a dark or reddish brown coloured oleoresin on tapping. The tapping is done by boring a hole about 2 cm diameters in tree of 1.5 m or more in girth. The hole reaches the pith about 0.9 m above ground and slopes downwards from the pith to the bark. The exudate is collected in a tin through a metallic lip. When flow stops, the hole is plugged with wood and the tree is given rest for 10 years. A healthy tree of 2.4 m girth yields 55 litres of oleoresin, the maximum yield reported being 180 litres. The oleoresin quality is inferior and has no commercial possibilities. It is used in preparation of wood varnishes.

Guggul (*Commiphora mukul)* is a shrub or small tree, found in the arid and semi-arid climate and is tolerant of poor soil. It is the source of fragrant gum-resin called Indian Bdellium/Indian Myrrh/Gum Guggul, obtained by making oblique incisions into the bark of the trunk at 7 to 10 cm apart with a pointed knife. A yellow milky juice oozes out from the incisions and darkness in colour on thickening. The resin is then scrapped off the wound with the knife. Each tree yields about 0.75 to 1 kg of resin. The resin has an aromatic taste and balsamic odour. Other Indian Commiphora spp such as *Commiphora roxburghii, Commiphora berryi* and *Commiphora caudata* species also yield minor quality of gum-resin. It is largely used as an incense as a fixative in perfumery, and as a substitute for African bdellium. It is used in medicine and is a substitute for myrrh and also used in pharmaceutical drug.

Harvesting and Extraction of Resin

The following commercial harvesting methods viz., Tapping and Blazes are employed for extraction of resins.

i) Selection of Trees

Resin tapping is performed in trees with more than 60 cm of girth at 1.30 m from the ground. Trees of and above 30 yr old should be tapped. Resin tapping is not recommended in trees younger than 30 years as resin yield is considerably lower in younger age.

ii) Tapping techniques in global level

a) **Chinese method:** A downward pointing V-shaped groove is cut every day, deep enough to reach the secondary xylem. The first groove is cut at 1.2 m above the ground and subsequent grooves are cut below it. The groove reaches roughly halfway around the tree circumference. No chemical stimulant is used. This method is used mainly in China.

b) **American method**: A horizontal groove is cut at every 15 to 18 days. The grooves are cut upward, the first at 20 cm above the ground. Only the bark and phloem are removed. The length of the grooves is about one third of the tree circumference and the height varies from 2 to 3 cm. A stimulant paste with 18 to 24 percent sulphuric acid (H_2SO_4) is applied. In the paste formulation, stimulants are also used as chemical adjuvants, such as CEPA

(2-chloroethyl-phosphonic acid, an ethylene precursor) or salicylic acid. This method is used in Brazil, Argentina, Portugal and Spain.

c) **Hugues or French method**: Slices of 8 to 10 cm wide is cut into the trunk at every 10 to 15 days, reaching the secondary xylem. The cut surface may extend to 1.8 m from the ground after two years of extraction. This method was developed in the mid-nineteenth century in France and is now used mainly in Indonesia.

d) **Mazek or Rill method:** V-shaped grooves, 2 to 3 mm wide are cut at every 3 to 7 days. The grooves are cut upwards. A stimulant of 50 percent H_2SO_4 and 50 percent hydrochloric acid is applied in the form of a spray. This method is currently used in Indonesia and India.

Tapping methods in India

In India, two kinds of tapping are in vogue, viz., a) Light continuous tapping b) Heavy tapping.

a) **Light continuous tapping:** Light continuous tapping is done in trees above 0.9m in girth. Trees between 0.9 and 1.8 m in girth are tapped in one channel, and those above 1.8 m in girth in two channels each at a time. The first channel or each set of two channels is tapped for five years. At the end of five years, a new channel or a new set of two channels as the case maybe, is started leaving an interspace of 10 cm between the old and new channels. At the end of the second five year period, another channel or set of two channels is again made leaving another interspace and so on till tapping has progressed channels, right around the tree. This is the common mode of tapping practiced in India and the trees are regularly tapped without any rest.

b) **Heavy tapping:** In heavy tapping (also known as tapping to death), maximum possible out turn of the resin is derived before the trees are due for felling and it is achieved by cutting on the tree as many channels as it can bear, with a minimum inter space of 10 cm between the successive channels. It is generally started five years in advance of the main felling in prospective regeneration areas and two years in advance of thinning in areas marked for thinning. The lowest girth prescribed for heavy tapping is 60 cm.

Tapping methods in Pine trees

The four methods of tapping in pine trees viz., Box method, Cup and lip method, Rill method and Bore hole method are commonly employed in India (Fig. 3.3).

Fig. 3.3 Resin tapping methods

a) **Box method:** This is the oldest method of resin tapping. A cavity or box of 10cm x 10 cm and upto 12 cm deep is chopped at the base of the trees. It is meant to collect the resin as it exudes from the blaze or incision that is made just below the box, by chipping bark and outer layer of the sapwood. The resin oozes out of the blaze and is collected in the box. However, this method would damage the trees and the trees tend to die within few years after resin tapping is started.

b) **Cup and lip method**: In this method, the outer bark of the tree is scraped off with the adze to a reasonably smooth surface of 60 cm long, 15 cm wide and 25 cm above the point where the lip is fixed. In light tapping, channels are initially located on south or south-west face of the tree as better yields are obtained from the warmer aspects. Subsequent channels are made in an anti-clockwise direction. A cut of about 15 cm broad and slightly slanting outwards is made with a curved chisel and mallet about 25 cm below the lower edge of the blaze. The lip, a rectangular piece of galvanized iron (15 cm x 5 cm), is driven into the cut to collect resin into a pot kept below. The pot is partially covered to prevent pieces of bark, dirt etc. from falling into the cone and also to minimize the evaporation of resins that accumulate into it. In order to open up the clogged resin ducts and aid in the smooth and continuous flow of resin, the channels are freshened at definite intervals. The cup and lip method of tapping has a number of disadvantages. Even though a channel depth of 2.5 cm has been prescribed for the blazes, very often the depth exceeds the prescribed limit. The inherent hacking action involved in the case of adze makes it very difficult to control the depth of the blazes. The tapper also makes much deeper blazes in the hope of getting more resin. Deep cuts around the hole results in loss of timber and make the trees less resistant to wind storms. Moreover, after covering the circumference of the tree, the second cycle of tapping is not possible because of the slow healing of deeper blazes. This generally results in an abrupt fall in resin production.

c) **Rill method**: This is an improved method, standardized at Forest Research Institute, Dehradun to overcome the disadvantages of the cup and lip method. In the rill method, the bark of the tree over a surface area of about 45 cm in height and 30 cm in width is removed with the help of a bark shaver. The surface is made very smooth and the thickness of the bark left should not be more than 2 mm to facilitate freshening of the blaze. The blaze frame is kept on the stem in the vertical portion, 15 cm above the ground level and the position of the blaze is marked with a marking gauge. The control groove is cut with a grove cutter by drawing it from top to bottom. The lip is then fixed in the tree with nails. For freshening of the blaze, the tapper stands near the tree on one side of the blaze and holds the freshening knife at the lowest point of the control groove. The knife is then pulled up by the tapper along with blaze line marked on the tree. The depth of the rill is about 2mm into the wood. After making a freshening on both arms of the blaze, a mixture of dilute sulphuric acid (20%) and dilute nitric acid (20%) at a ratio of 1:1 is sprayed on the freshly cut rill with the help of a spray bottle. Exudation of oleoresin starts soon after the rills are made. The pot containing the oleoresin is emptied into a collection can. The resin adhering to the pot is removed with the help of a scraper. The control groove is also increased to avoid the accumulation of resin in it

d) **Bore hole method**: Forest Research Institute, Dehradun has developed a new tapping technique known as the borehole technique of resin extraction from *Pinus roxburghii*. In this method, holes are made near the ground level with the help of a machine into tree sapwood to open the resin ducts and exudating resin is collected in a closed container. The hole in each tree is done approximately 10 cm above the ground. It has been found that holes of 15cm depth and 2.5 cm diameter are suitable for obtaining maximum resin yield. The holes are drilled straight into the tree stem with a slight slope towards the opening so that resin drains freely. Immediately after making the hole, the stimulants/ chemicals (mixture of sulphuric acid and ethephon i.e., 2-chloro-ethyl phosphonic acid) are sprayed inside each freshly made hole. A spray volume of 1 to 2 ml is applied to each hole. Chemical treatment is done once only, immediately after boring holes. After treatment, a spout is installed inside the hole by gently hammering with a small mallet or pushing with palm of the hand to achieve compression fitting in the hole. The spout is meant for joining the collection container (polybag made up of high density poly ethylene, HDPE, 35.3 x 12 cm) tightened to each spout. Once the polybag is filled with resin, it is removed from the tree and poured into a collection can and immediately a new polybag is tied for future collection of resin. The advantages of bore hole method of resin extraction are as follows. There is less stress due to small size (2.5 cm diameter) of the hole. The hole heals fast. The technique is very suitable for the protection of tree against fire, insect, pest and diseases. Prolong resin flow can be obtained from bore hole for a period of several months without wounding the stem. The holes are made with the help of a machine; therefore, the labour productivity of the technique is several folds greater than other method. The technique could be very effective in conservation and management of pine resources.

Processing and Value addition of Resins

Resin is collected in bucket and brought to central places from where it is taken to the factory. Resin obtained from trees is distilled into turpentine oil and rosin. Crude resin is in a very impure state when it reaches the factory. It contains water, bark, needles, dirt and even insects. They are stored in a suitable shed after being brought to the factory. Before processing the top of the tin bucket is cut and the dirt that is floating on the top of the resin is removed manually. During winter, resin is a bit hard to work, so some amount of turpentine oil is mixed to dilute the resin. The screw elevator lifts this resin along with the impurities to the mixing vats which are large vessels made of mild steel. They have V-shaped bottoms that are provided with steam jackets for heating the resin. The resin is heated indirectly in this vat by passing steam through copper coils for 4 to 6 hours to melt the resin in it. The temperature molten resin is then stirred with iron arms mechanically and passed through 40 mesh stainless steel wire net to remove bigger impurities like pine needles, bark pieces, etc. and then pumped by centrifugal pump into a rest vat.

The resin is allowed to rest in the rest vat for about 18 to 20 hours. During this rest period, the impurities and dust if present, settle down by gravity. The dust and other impurities are removed through a bottom valve and sent for processing separately as inferior grades of rosin. Resin from rest vat is carried to the compression chamber, from which it is carried to distillation kettle by steam pressure. In the distillation kettle, the resin is cooked for 1-2 hours by indirect steam. The molten resin remains inside the copper tubes and steam remains in the outside. In this process the turpentine oil and moisture present in the resin get evaporated and condense in the condenser. Turpentine oil is collected in a separator. The rosin left in the distillation kettle is taken out at 165°C and collected in trolleys, wherein about 100-200g of oxalic acid (depending on the quality of resin) is mixed to increase the transparency or shine of rosin (Fig. 3.4 and Fig. 3.5).

Fig. 3.4 Rosin

Fig. 3.5 Turpentine

Storage of Rosin

Rosin is packed in Tin Patra Barrels of 200 kg of capacity and stored in cool condition under shed. Both rosin and turpentine oil find wide use in various industries. They are graded into 12 categories for commercial use.

Properties of Rosin

- Rosin is brittle solid and consists mainly of free organic acids called rosin/resin acid and related neutral materials like resin acid esters, anhydrides, aldehydes, hydrocarbons, unsaponifiable matter and fatty acids.
- Its colour varies from light yellow to red, brown, blue or black.
- Its specific gravity is 1.08 and melting point varies from 100 to 140°C.

Applications of Rosin

- Rosin is used in making cement varnishes, paints, sealing wax, adhesives, inks and soaps.
- It is used as a dressing for machine belts and bows of violins and cellos.
- It is used in the preparation of certain metals for soldering.
- Rosin gives a stiff coated surface to certain kinds of paper. This is needed for printing and writing.
- In pharmacy, rosin is used in some plasters and similar ointments preparations.
- Athletes rub it upon their hands or the soles of their shoes to prevent slipping.
- It is also extensively used for its friction increasing capacity in stringed instruments.
- The tin lead solder commonly used in electronics has about 1% of rosin as a flux.
- It is also used in the manufacturing of linolium, sealing wax, oil, cloth, special flooring compounds and covering lubricating compounds and disinfectants.

Turpentine

- Turpentine is a fluid obtained by the distillation of resin harvested from living trees, mainly pines.
- Turpentine is composed of terpenes, primarily the monoterpenes, alpha and beta-pinene with lesser amount of carene, camphene, dipentene and terpinolene.
- Mainly used as a specialized solvent
- It is also a source of material for organic syntheses
- Turpentine is composed of terpenes, primarily the monoterpenes, alpha- and beta-pinene, with lesser amounts of carene, camphene, dipentene and terpinolene.
- It is used in food and beverages
- Distilled turpentine oil is used as a flavoring ingredient.
- Turpentine oil is used in soap and cosmetics and also as a paint solvent.
- It is also added to perfumes, foods and cleaning agents as a fragrance.

Suggested Readings

Farooqi MIH 2008 *Dictionary of Indian Plant Gums, Resins, Dyes and Related Products,* Lucknow: Sidrah Publishers

Hosakatte Niranjana Murthy 2022 *Gums, Resins and Latexes of Plant Origin: Chemistry, Biological Activities and Uses,* Switzerland AG: Springer Nature

Howes FN 2001 *Vegetable Gums and Resins,* Jodhpur: Scientific publishers

Nussinovitch A 2009 Plant gum exudates of the world: sources, distribution, properties, and applications. CRC Press.

Pratibha Bhatnagar 2014 *Gums and Resin Yielding Plants,* Jaipur: Pointer Publishers

Rizvi S S 2014 Thermodynamic properties of foods in dehydration. In Engineering properties of foods. CRC Press, Pp. 261-348.

Sanchez C D Renard Robert C Schmitt and J Lefebvre 2002 Structure and rheological properties of acacia gum dispersions. Food hydrocolloids, 16(3): 257-267.

CHAPTER 4
Tannins

M.P. Divya and C.N. Hari Prasath
Tamil Nadu Agricultural University, Coimbatore

Tannins are secretion products found in plant tissues in small or large quantities. These are simple chemical compounds of carbon, hydrogen and oxygen along with some nitrogen. The name tanning is given to organic substances which have the property of combining with the albumen and gelatin to form an insoluble compound which will resist decay. By their action on animal hides and skins, they cause changes which make the hides and skins resistant to decomposition, make them flexible, strong and greatly improved in wearing qualities. Hides and skins so treated are called leather. The vegetable tannins have held their own to a large extent, as synthetic tannins cannot be made cheap to compete with natural sources. As secondary metabolites, tannins are sequestered in vacuoles within the plant cell, which protects the other cell components. Tannins are polyphenolic substances, yellowish or brownish bitter tasting organic substances present in some galls, barks and other plant tissues, consisting of derivatives of gallic acid and amorphous, astringent and feebly acidic in the nature. It is water soluble nature allows easy extraction with molecular weight in the range of 500-3000 $gmol^{-1}$ and is useful in various applications in the chemical and pharmaceutical industries. The term tannin was introduced by Seguin in the year 1976 to indicate various plants extracts which can convert hides and skins into leather. In this process, tannin molecules cross-link the protein and make it more resistant to bacterial and fungal attacks. Tannin is a simple chemical compound made of carbon, hydrogen and oxygen. Tannins usually are located in the vacuoles of plant cells and many cases, make up a sizable proportion of the dry weight of plants. Tannins often are concentrated in epidermal tissues and the bark of woody plants but these phenolic compounds may be found in leaves, roots, stems, fruits, bark, wood, or other plant parts. Tannin content may vary widely during the growing season in a particular plant part.

Structure of Tannin

Tannin is a large polyphenolic compound containing sufficient hydroxyls and other suitable groups (such as carboxyls) to form strong complexes with proteins and other macromolecules. It is obtained from plants like Wattle (*Acacia* spp.), Oak (*Quercus* spp.), Eucalyptus (*Eucalyptus* spp.), Birch (*Betula* spp.), Willow (*Salix caprea*), Pine (*Pinus* spp.). It is found in abundance in the tree bark, wood, fruit, leaves and roots and also in plant gall. They are found as shapeless yellowish or light brown masses like powder, flakes or sponges. They are incompatible with alkalis, gelatin, heavy metals, iron, lime water, metallic salts, strong oxidizing agents and zinc sulfate.

Occurrence

Tannins are distributed all over the plant kingdom. Tannins are formed upon the decomposition of vegetation. These compounds are large molecular weight organics that fall under two large primary categories: humic acids and fulvic acids. These are simplified structures but they show the complexity of tannins. Gallic acid, guaiacyl, syringyl, and cresylic moieties are the building blocks of these large organic molecules. The structure of the tannin varies depending on the plant life in a given area. Tannins are found in leaf tissues, bud tissues, seed tissues, root tissues and stem tissues. An example of the location of the tannins in the stem tissue is that they are often found in the growth areas of trees, such as the secondary phloem and xylem and the layer between the cortex and epidermis. Tannins may help regulate the growth of these tissues. They are also found in the heartwood of conifers and may play a role in inhibiting microbial activity, thus resulting in the natural durability of the wood. Tannins can cause a yellow to brown cast in water and may also impart a taste and odour. They will generally be found in surface water supplies or shallow wells. Although these compounds are not a health risk, they are aesthetically displeasing. At present, tannins are difficult to remove from water. Tannins can be removed by various methods. One treatment method may be effective in one area but may be very ineffective 10 miles down the road. It is all dependent on the vegetation in a given area. Styrene based macroporous anion resin has long been used to remove tannins from water. This type of resin has worked well in some areas but poorly in others. More recently, acrylic based resins have emerged on the market and are producing better results when removing tannins from water. The acrylics can also be manufactured with a macroporous structure. The additional porosity enables the tannins to be regenerated from the resin more effectively.

Tannins in Plant Tissues

Bud tissues: Most common in the outer part of the bud, probably as protection against freezing.

Leaf tissues: Most common in the upper epidermis. However, in evergreen plants, tannins are evenly distributed in all leaf tissues. They serve to reduce palatability and thus, protect against predators.

Root tissues: Most common in the hypodermis (just below the epidermis). They probably act as a chemical barrier to penetration and colonization of roots by plant pathogens.

Seed tissues: They are located mainly in a layer between the outer integument and the aleuronic layer. They have been associated with the maintenance of plant dormancy and have allopathic and bactericidal properties.

Stem tissues: They are often found in the active growth areas of the trees such as the secondary phloem and xylem and the layer between the epidermis and cortex. Tannins may have a role in the growth regulation of these tissues. They are also found in the heartwood of conifers and may contribute to the natural durability of wood by inhibiting microbial activity.

Properties of Tannin

i) Physical properties of tannin: Tannins are non-crystallizable compounds. They are soluble in water forming colloidal solutions with acidic reaction and sharp astringent taste. They are soluble in dilute alkalis, alcohol, glycerol and acetone but only sparingly soluble in other organic solvents. Their solutions precipitate heavy metals, alkaloids, glycosides and protein.

ii) Medicinal and biological properties: Tannin containing drugs precipitate proteins and have been traditionally used as styptics (stop hemorrhage) and internally for the protection of inflamed surfaces of the mouth and throat. They play an important role in the treatment of burns. They form a mild antiseptic protective layer on the surface of the injured skin below which regeneration of new tissue takes place. They act as anti-diarrheal, although not recommended in this respect as they usually delay the elimination of bacterial toxins from the body. Tannins have been employed as an antidote for poisoning by heavy metals, alkaloids and certain glycosides due to their precipitation as tannates. Recently tannins as most polyphenols are proved to have a potent antioxidant effect.

Classification of Tannin

a) Types of tannin based on structure

There are three structural types of tannins viz., Hydrolyzable tannins, Condensed tannins and Pseudo tannins.

i) Hydrolyzable tannins

These tannins are hydrolyzed by acids or enzymes quickly and the products of hydrolysis are gallic acid or ellagic acid. Complex esters of gallic acid with carbohydrate (usually glucose) called gallotannins, are widely distributed among dicotyledonous plants. The most common gallotannin is β-pentagalloyl glucose. Hydrolyzable tannins are hydrolyzed by weak acids or weak bases to produce carbohydrate and phenolic acids. Examples of gallotannins are the gallic acid esters of glucose in tannic acid, found in the leaves and bark of many plant species. Ellagitannins are hydrolyzable tannin and they often co-occur with gallotannins especially in oak, eucalyptus, chestnut and some *Caesalpinia* spp., and *Terminalia chebula*.

Hydrolyzable tannins are abundant in the leaves, fruits, pods and galls (sometimes bark and wood) of a number of dicotyledonous families such as Aceraceae, Anacardiaceae, Combretaceae, Dilleniaceae, Dipterocarpaceae, Droseraceae, Ebenaceae, Ericaceae, Euphorbiaceae, Fabaceae (Caesalpinioideae), Fagaceae, Fouquieriaceae, Geraniaceae, Haloragaceae, Hamamelidaceae Juglandaceae, Lecythidaceae, Loranthaceae, Lythraceae, Melastomataceae, Myricaceae, Myrtaceae, Nymphaeaceae, Nyssaceae, Onagraceae, Plumbaginaceae, Polygonaceae, Rhizophoraceae, Rosaceae, Santalaceae, Simaroubaceae and Paeoniaceae. The major sources of commercial ellagitannins are myrobalans, divi-divi, algabobilla, valonia and the bark of oak and spanish chestnut.

ii) Condensed tannins

They are called as nonhydrolyzable tannins, phlobatannins, or proanthocyanidins. They are much resistant to hydrolysis. They are related to flavonoid pigments because they are formed via derivatives of flavones, like catechin or flavan-3-ol or flavan-3,4-diols. Unlike the hydrolyzable tannins, on treatment with enzymes or mineral acids, they are polymerized or decomposed into red colored substances called phlobaphenes. They are insoluble in water and indicate the typical brownish red colour of many plants and drugs colour. On dry distillation, they yield catechol. Tannins with ferric chloride produce brownish green. They are distributed in different parts of plants. The green tea and hamamelis leaves, cinchona, cinnamon and wild cherry bark and pale and black catechu contain these types of tannins.

iii) Pseudo tannins

This is not such a separate group of tannins, but may be treated as a subgroup because they do not obey to gold-beaters skin test. Pseudo tannins are low molecular weight compounds associated with other compounds. Chlorogenic acid in coffee and *Strychnos nux-vomica*, ipecacuanha acid in *Carapichea ipecacuanha* and catechins in cocoa are the examples of pseudo tannins. The detection test for chlorogenic acid is carried out by extracting the drug with water.

b) Types of tannin based on the presence

Tannin is obtained from different parts of plants. Based on their presence, these are classified into Wood tans, Bark tans, Fruit tans and Leaf tans (Fig. 4.1)

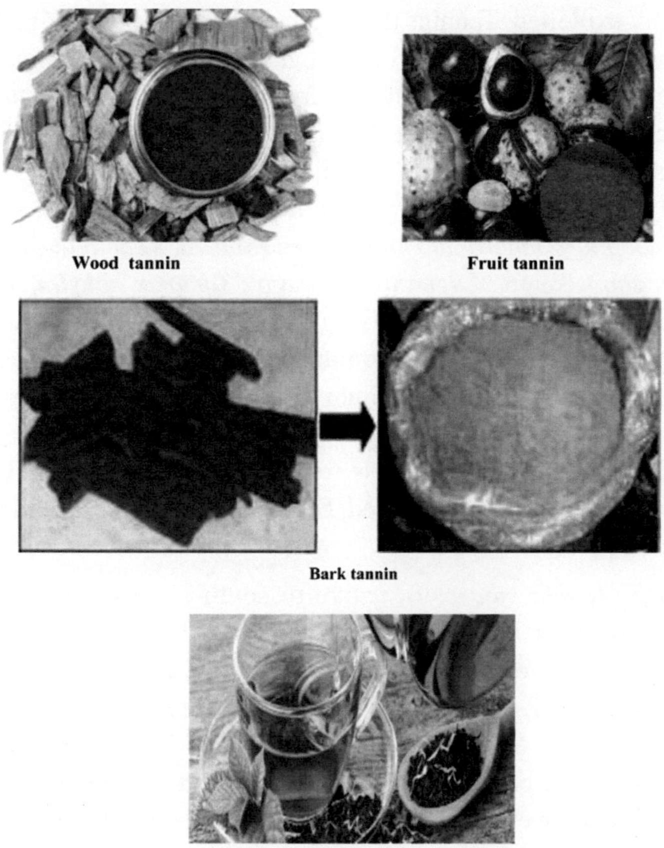

Fig. 4.1 Tannin from different plant parts

i) Wood tans

Tree wood yielding tanning materials are known as wood tans. The heartwood contains 20-27 percent tannin, which is obtained by cutting the wood into small chips and extracting the tannin with water. In India, the cutch obtained from Khair (*Acacia catechu*) wood is used for tannin purposes. *Acacia catechu* is a characteristic tree of Khair-sissoo forests, southern thorn forests, ravine thorn forests and very dry teak forests. It is adaptable to a wide range of well drained soils, even poor stony types which are unsuitable for many other trees. The tannin content of *Acacia catechu* is 57 to 59 percent. Cutch is used as a blend along with wattle, it is used for producing heavy leather. It is also used for dyeing leathers.

ii) Bark tans

The bark of several tree species yields a good amount of tannins and which are commercially exploited. Tannins that become stored in the bark of trees protect the tree from being infected by bacteria or fungi. In this case, the tannins precipitate out the enzymes and other protein exudates from bacteria and fungi, thus not allowing these organisms to infect the tree. Most of the tan yield trees are naturally found in the forest over a large area. The major bark tan yield tree species are *Acacia mearnsii, Acacia nilotica, Cassia auriculata, Cassia fistula, Ceriops roxburghiana, C.tagal, Rhizophora mucronata* and *Shorea robusta, Acacia leucophloea, Bridelia retusa, Castanopsis* spp., *Lagerstroemia* spp., *Tamarix aphylla, Terminalia alata* and *Quercus* spp. Generally, the fast growing tree species contain more bark tan compared to slow growing trees. Commonly, the beginning of the growing season is considered to be the most optimum period for harvesting bark for collecting tans because tannin content is highest during that period. With help of debarking axes or debarking spade, the bark is removed in from the tree. Then the bark is collected, dried, stored and marketed. Some of the bark tan yield tree species are discussed below:

Black wattle *(Acacia mearnsii)*, native to south-eastern Australia and has been successfully growing in high altitude regions of India. It is cultivated in many parts of the country for its tans. In India, it is commonly grown in Kodaikanal, Ooty hills, Kohima, Kerala, Meghalaya and hill regions of Jammu and Kashmir. In some parts of the world, *A. mearnsii* is considered to be an invasive species because of the large production of seeds each year. In India, *A. mearnsii* yields 5000 tons of tannins every year. The tannins produced by the wattle bark are of exceptional quality with an average yield of 35 to 45 percent from green bark. The colour of the wattle tannin is yellow or red. The wattle tan is an astringent and also a catechol type of tannin. The concentration of the wattle tan is liquor, this will improve the shelf life of tannin i.e., on standing, liquor does not lose its stability. The wattle tan is mainly used in leather industries and plywood industries in adhesives formulations.

Black babul *(Acacia nilotica)* is a small medium sized tree found in the drier tracts of northern and southern tropical thorn forest and southern dry mixed deciduous forest. The bark of babul is used as a tanning material on a large scale. The babul tree is commonly growing in tank beds and homesteads in rural areas in India. In India, the estimated quantity of babul tan is about 1 lakh tonnes and is majorly consumed by tanneries in Kanpur. In babul, the tannin content in bark varies accordingly to the age. The older age trees yield more tannin compared to younger trees. The tannin content in the bark of babul is 7 to 12 %. The deep color and high non-tannin are the major disadvantages of this bark tan but babul tan is considered to be good tannin for heavy leather because it gives dark coloured, durable and firm leather. Blended with myrobalan tan gives better finished leather. Babul tan is not considered as good for kips and half tanned hides.

Avaram *(Cassia auriculata)* is a medium sized shrub, growing wild in the Deccan plateau and parts of south India. The bark of this species is largely used as a tanning material in parts of South India and also some parts of Rajasthan and Maharashtra. The bark of the avaram is considered one of the best known tanning materials in India because the bark is easy to use, tannin loss is very low, pale coloured leather with good strength and yields uniformly successful results. In India, the estimated annual production of avaram tan is 1.25 lakh tonnes. The tannin content of avaram varies from age, the older age plants yield more tannin content up to 23 percent, while the young shrubs yield around 12% or sometimes even less. The method of collection of bark consists of cutting shoots from the base and stripping off the bark from shoots. The bark obtained from shoots is dried and transported to tanneries.

Indian laburnum (*Cassia fistula*) is native to the Indian subcontinent and it is grown in tropical southern and northern thorn forests and tropical moist forests. It is common among the tanneries in the southern part of Tamil Nadu. The tan percent of this tree is around 13-15%. It produces smooth grained and pale leather. It is used in the tannage of kips and lightweight hides. *C. fistula* bark is sometimes used as a substitute for avaram bark in blends. This species of tan cannot be used directly because this tree possesses low penetration power.

Ceriops roxburghiana is a shrub or small evergreen mangrove tree. It grows predominantly in mangrove areas of Sundarbans, Tamil Nadu, Gujarat, Maharashtra, Andaman and the Nicobar Islands. This species contains valuable tanning material in different plant parts majorly in bark and leaves. The bark contains 20 to 37% tannin and the leaves contain 9 to 15% tannin. The bark is an astringent, so when used it imparts red colour to the leather and is mostly used for manufacturing heavy leathers. The important characteristics of this mangrove liquor are its stability and low loss of tannin standing.

Shorea robusta is large quantities of the bark of sal (*Shorea robusta*) become available during fellings in the forest. The bark contains 3-9% tannin and furnishes a very tough and reddish color to the leather.

Terminalia arjuna is a trcc occurring throughout India usually on banks of rivers and streams, especially in central and southern India. The tannin content in the bark is 20-24%. The color of leather produced is light-brown.

Rhizophora mucronata is a well-known mangrove tree whose bark is rich in tannin, but it contains an excessive quantity of coloring matter. It is used in local tanneries.

iii) Fruit tan

Fruit of some tree species is employed in tanning industries for extraction of different tannins. The important tan yield species are discussed below.

Acacia nilotica pods of babul produce good quality tannin. The average tannin percentage is 18 to 27 % in the pods after the seeds have been removed from the pods. It gives a good colour to the leather and has a softening effect. A large quantity of babul fruit tanning is used in Maharashtra, Madhya Pradesh, Uttarakhand and Uttar Pradesh. This tan is not only used for tannin purposes and also to soften and impart a good colour to the leather.

Terminalia chebula, commonly known as myrobalan, grown in greater parts of India, particularly in sal and teak mixed forests and mixed deciduous forests. It is mainly growing in Madhya Pradesh, Maharashtra, Tamil Nadu, Odisha, Karnataka, Andhra Pradesh, Himachal Pradesh and Bihar. Myrobalan is onc of the important fruit source tannins in India. The fruits of myrobalan are collected during December to February. The estimated annual yield is nearly 50,000 tonnes. The tannin content varies from 12 to 48% with an average of 34%. It is hydrolyzable tannin and so it finds restricted to use in the leather industry. Generally, it is used in the tanning industries for improving colour fastness. Myrobalan tannin is not suitable for self-tanning material but it is used as a blend along with other tanning materials because it has low penetration power; it produces soft, mellow and spongy leather which lacks good wearing properties.

Caesalpinia coriaria is commonly called divi- divi. It is a small tree, native to central and South Australia. In India, it has been successfully planted in parts of Tamil Nadu and Maharashtra. The average tannin content is nearly 38%. The important characteristics of pods are high tannin content and easy tannin extraction process. Divi-divi tannin is not used directly because in moisture conditions, the leather becomes soft and spongy and in dry weather, it loses its pliability.

Anacardium occidentale (cashew nut) is native to South Africa, but it has been successfully cultivated in many parts of India, particularly found in the coastal regions of Andhra Pradesh and some parts of Tamil Nadu. The tannin obtained from the testa of cashew nuts is of catechol type. The average tannin content of cashew nut is around 23%. The testa of the species is rich in tannin content and can be used for making tannin extract. This tannin has high penetrating power and yields tough leather with a red tint.

Shorea robusta is commonly known as sal, found in central and northern Indian regions. Commonly found in the state of Uttarakhand, Uttar Pradesh, Madhya Pradesh, Bihar, Odisha, West Bengal and Assam. The tannin obtained from sal fruit is an ellagitannin class of the hydrolyzable type. The average tannin content in the original and deoiled seeds is 7.2 and 9.6% respectively. The corilagin, chebulinic acid, gallic acid and ellagic acid of phenolic constituents are present in the sal seeds. The seed tannin can be used as a blend with babul. This tannin cannot be used as a self-tanning material.

Zizyphus xylopyrus (Gothber) is a tan fruit of importance. Its fruits impart a black colour to the leather.

Emblica officinalis fruits are rich in tannin. Immature fruits are usually employed in combination with other tan stuff such as myrobalans.

iv) Leaf tan

Leaf of some of the plants provides tanning material for local uses. Generally, village artisans and shoemakers use leaves for tanning leather. They are not used extensively for tanning purposes by large tanneries. Major leaf tanning producing tree species are *Anogeissus latifolia, Carissa spinarum, Continus coggygria, Emblica officinalis* and *Lawsonia inermis*. In some species, leaf gall also produces the tan and the major species are *Tamarix* spp., *Garuga pinnata, Prosopis cineraria, Terminalia* spp., etc.

Anogeissus latifolia is commonly known as axle wood and is found almost throughout the sub-Himalayan tract, Bihar, Chhota Nagpur and central and southern parts of India. A mixture of green leaves, red leaves and petiole contains nearly 38.5% of tannin. Compare to old leaves, young leaves are producing maximum tannin content up to 55%. This tannin is hydrolyzable and finds limited uses in the leather industry and the leather produced from axle wood tan is highly sensitive to light.

Carissa spinarum commonly grown in scrub jungles of southern India, central India and the foothills of Shiwalik regions. Its leaves contain upto 9 to 15% tannin.

It is used in combination with other tannins such as twig bark of *Emblica officinalis*. The tannage is slow and causes swelling in the hide.

Emblica officinalis is commonly found in the dry deciduous forests throughout India. The leaves of this tree yield tannin generally used in local tanneries. The average tannin content in leaves is around 22 %.

Tannin Identification Test

a) Gelatin test

An aqueous solution of 1% gelatin and 10% sodium chloride is added to a solution of tannin. If a white buff coloured precipitate forms which confirm the presence of tannins and pseudo tannins.

b) Gold beaters skin test

A small piece of gold beater skin (membrane prepared from the intestine of an ox) is soaked in 20% hydrochloric acid, rinsed with distilled water and placed in a solution of tannin for 5 min. The skin piece is washed with distilled water and kept in a 1% solution of ferrous sulphate. The production of brown or black color on the skin indicates the presence of tannins.

c) Phenazone test

10 ml of aqueous extract of tannin and sodium acid phosphate is heated and cooled and filtered. About 2% solution of phenazone is added to the filtrate. The formation of a bulky colored precipitate shows the presence of tannin.

d) Match stick test (Catechin test)

A match stick is dipped in aqueous plant extract, dried near a burner and moistened with concentrated hydrochloric acid. On warming near the flame, the matchstick wood turns pink or red due to the formation of phloroglucinol which confirms the presence of tannin.

e) Test with ferric chloride

To the solution of tannin, ferric chloride solution is added. A blue, black, violet or green colour precipitate confirms the presence of tannins.

Tannin Extraction Methods

i) Maceration method

Maceration is one of the techniques used for tannin extraction from medicinal plants. Maceration is the simplest technique of extraction where the plant powder is placed in a closed vessel and soaked with the corresponding amount of solvent for a specified period of time until the tannins are dissolved in the solvent. In the first stage of the maceration, osmosis occurs before diffusion occurs, in later stages, osmosis occurs simultaneously with diffusion. The first stage is usually very short and it can be a few seconds or hundredth second. The first-stage time depends on the type of solvent, the extraction material and the extraction conditions. The technique is considered as simple and low cost way to extract tannins from medicinal plants. This technique is popular choice for researchers because of non-complicated utensil and equipment. However, the duration of extraction time is long, the ratio of solvent to material is higher, the extraction efficiency is low and the volume of extraction tanks and the factory area are wide compared to other techniques. The yield of tannin extraction increases when hydrogen bonds between tannins and proteins are broken or interactions between tannins and proteins during extraction are not formed.

The yield of tannin extraction depends on many factors, for example, type of solvent, plant species, temperature and time during the extraction process and pressure in vessel. In the maceration, temperature of extraction can be hot, cold or warm due to tannin characters and sample. In cold extraction, it requires thermolabile compounds. The hot extraction technique is recommended for high temperature leading to decomposition proteins and cell structure during heating. When tannin extraction from *Galium tunetanum*, the acetone being a better solvent, compared to ethanol (30%), Na_2SO_3 solutions and NaOH solution. When extracting tannins from *Acacia mollissima* barks, methanol is the best solvent at room temperature and ethanol exhibits as the best solvent at 60°C. Less tannin is extracted with aqueous or acidic methanol than with aqueous acetone. For fresh barks of *Acacia mangium*, 50% acetone brings high yield of tannin extract. The choice of the solvent types and the extraction temperature depends on the properties of the tannin in each plant. For extracting tannins from seaweed, ethanol is the best solvent because of high purified tannins and good activity.

ii) Method of Decoction

The decoction is similar to maceration method. However, the mixture is continuously heated at 100°C during decoction. Reaction kinetics due to the heat is stronger than the method of maceration. For extracting condensed tannins, the decoction is more effective than the maceration. In these methods, condensed tannin concentration depends on polarity of the extraction solvent. The mixture is then cooled to room temperature and the filtrate collection. Aqueous is a good solvent for the extraction of condensed tannins by decoction and maceration. The decoction is very much suitable for non-destroyed substances by the heat.

In addition to diffusion pressure and osmosis pressure, tannins are also separated from the material by the energy supplied from the thermal energy of the solvent and the affinity of the solvent. The thermal energy of the solvent separate the tannins from the cells through the action on the bonds between tannins and cells. The affinity of the solvent will compete and pull tannins out of the material. The method requires easy and cheap equipment. It is suitable for all production scales. However, the method consumes more energy than the maceration and it is not suitable for the extraction of heat sensitive tannins.

iii) Pressurized Water Extraction

The method is similar to infusion, hydrodistillation, decoction and maceration. However, high pressure from 100 to 150 bars is used in this method. Extraction temperature ranged from 60 to 100°C for 30 min static extraction. The solvent of water is commonly used in the method and the other method includes subcritical water extraction, superheated water extraction, pressurized liquid extraction and accelerated solvent extraction.

Pressurized water extraction is divided into static pressurized water extraction and dynamic pressurized water extraction. The residence time of tannins in dynamic pressurized water extraction is shorter than in static pressurized water extraction and tannins are less degraded in dynamic pressurized water extraction. Hence, the extraction efficiency and the degradation of tannins depend on the extraction time. The distribution ratio of tannins into the water plays an important role in the efficiency of tannin extraction. Dynamic pressurized water extraction is a higher investment which is more difficult to use than static pressurized water extraction. The temperature of extraction is often from the atmospheric boiling point of water (100°C / 273 K, 0.1 MPa) to the critical point of water (374°C /647 K, 22.1 MPa).

iv) Ultrasound Method

The method of ultrasound is ultrasonic-assisted maceration method commonly used for the extraction of tannins from plants, microalgae and seaweeds, for instance, condensed tannins, hydrolysable tannins and valonea tannins. In this method, ultrasonic power, temperature and time of ultrasound are the factors that strongly influence the yield of tannin extraction. The tannins obtained by ultrasonic-assisted maceration method are 17.6% higher than traditional methods. In this method, the efficiency also depends on sound waves, which produce the mechanical vibrations in the solid causing implosion of cavitation bubbles and formed sheer forces. Thus, cell structure of plant is destroyed and the yield of tannin extraction is highly increased. The acoustic waves are often most commonly measured more than >20 kHz. Ultrasound-assisted extraction is the technology of low investment and high efficiency for the extraction of tannins. Ultrasound assisted extraction method leads to efficient improvement of mass transfer and tannin extraction yield, reducing consumption of time and solvent compared to the conventional method. Ultrasound-assisted extraction method is environmental friendly method, less risk of chemistry and physics and less impact on molecular structural properties of tannins in plants (Fig. 4.2).

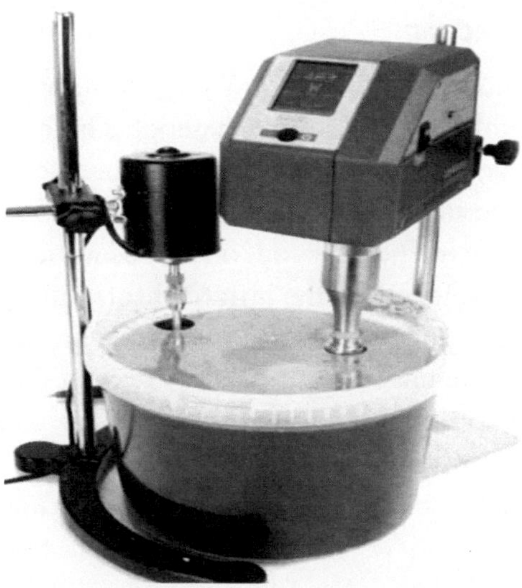

Fig. 4.2 Ultrasound Method of tannin extraction

v) Microwave-Assisted Extraction

The microwave-assisted extraction is a good choice for the extraction of tannins using microwave energy (electromagnetic radiations in the frequency range from 300 MHz to 300 GHz). The heating mechanism of microwave is very special. Ionic conduction and dipole rotation cause the transformation of the microwave energy to heat through interacting with polar molecules. The microwave supplies the high temperature (100 -150°C) in the extraction process. The yield of tannin extraction strongly depends on the penetration depth of microwaves into plants and the power of microwaves. Inside the dielectric constant, the moisture content and temperature of plants and the frequency of the electrical field decide the penetration of microwave depth. The water in plants absorbs microwave energy to reach the superheated state and cell structure disruption.

The solvents such as water, ethanol and methanol strongly absorb microwaves energy and their dielectric constant is high compared to toluene or hexane. Toluene and hexane maintain the surrounding extraction solvent in the cold state because of their dielectric constant. Water, ethanol and methanol are advised for the extraction of tannins, because tannins are very soluble. Microwave-assisted maceration is a simple, time-saving method, less cost of solvent and energy. The microwave technique brings the higher efficiency of tannin extraction than ultrasound method. The yield of tannin extraction in microwave method is 1.25 times more than maceration method.

vi) Infrared-Assisted Extraction

Infrared-assisted extraction is a maceration using an infrared lamp to supply the heat and the penetration of infrared into the material is higher than other methods. In conventional methods, the solvent is heated before heating the material. In infrared method, the material is directly heated without heating the solvent. The effectiveness of the infrared method depends on the absorption characteristics of the extracting solvent, the wavelength of the infrared heater and the distance between the lamp and the material. Electromagnetic waves excite tannins of material in the modes of twisting, stretching and bending, increasing the efficiency of tannin extraction. Infrared method exhibits a higher extraction yield than microwave-assisted extraction and ultrasound extraction. Infrared-assisted extraction is low cost, but extraction efficiency of tannins from plants is high and the time of infrared extraction is about 30 min.

vii) Soxhlet Extraction

Extraction method and solvent play an important role in the extraction of tannins. Soxhlet extraction is a good method when compared to cold maceration. The effectiveness of soxhlet extraction is mainly based on the evaporation temperature and polarity index of the solvent. The suitable temperature of extraction improves the diffusion ratio of tannins into the solvent and the circulation of the solvent. Fifty per cent ethanol gives the highest yield of tannins from herbal plants. Soxhlet extraction with water content is also evaluated as the best method for gallic and ellagic acid extractions. Solvent polarity, solvent-to-solid ratio and contact time significantly affect the yield of tannin extraction. The solvents such as water or ethanol-water mixtures are used for the extraction of the active hydrolysable tannins. However, soxhlet method needs long duration and high amount of solvent. When the evaporation temperature of the solvents is high, thermal destruction of tannins will occur. Extraction time ranged from 3 h or 6 h due to the materials (Fig. 4.3).

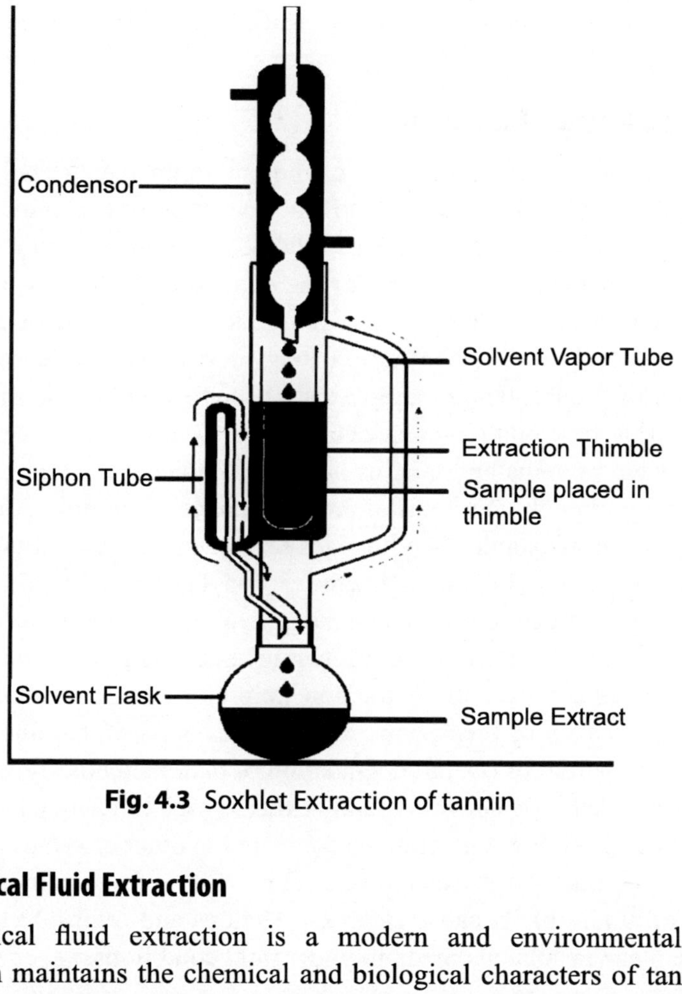

Fig. 4.3 Soxhlet Extraction of tannin

viii) Supercritical Fluid Extraction

The supercritical fluid extraction is a modern and environmental friendly method, which maintains the chemical and biological characters of tannins. The supercritical solvents have an important role in the efficiency of tannin extraction. The supercritical solvents are used as extraction solvent in the method including carbon dioxide, hexane, pentane, butane, nitrous oxide, sulfur hexafluoride and fluorinated hydrocarbons. Supercritical carbon dioxide is often the choice because of its low critical pressure (100 and 450 bar). However, the selectivity of CO_2 is not good.

For improving the problems, a co-solvent or modifier is combined with CO_2 in tannin extraction. Water and ethanol-water cosolvents exhibit a good result in the separation of the less polar compounds and the hydrolysable tannins. The diffusion coefficient of the supercritical fluid is higher than the liquid solvent, but the viscosity and the surface tension of the supercritical fluid are lower than the liquid solvent. Hence, the diffusion and mass transfer of tannins are better when using the supercritical fluid. The solvation power of supercritical fluid can be changed via the change of temperature and/or pressure.

ix) Enzyme Assisted Extraction

Enzyme-assisted extraction is one of the environmental friendly methods using an enzyme or complex enzymes to improve bioactive extraction yield via the disruption of the material cell. This method is combined with various methods, but high temperature is not used for the extraction of tannins because protein is precipitated under high temperature. Extraction time or time of enzyme treatment is sufficient and not long because a long time leads to release tannins and the formation of protein-tannin complexes. Of course, not all tannins precipitate all proteins, this precipitation is selective. The tannin-protein complexes are only formed when proteins and tannins have high molecular weight, an open flexible structure, hydrophobic proteins of proline richness. The linkages between tannins and proteins are unstable. The linkages between proteins and tannins are formed from phenolic groups of tannins and the carbonyl groups of peptides. For enhancing the extraction effectiveness, the treatment of the material cell by the enzyme complex should be used prior to tannin extraction. Enzymatic pre-treatment achieves a reduction in the extraction time and energy consumption. Enzymes complexes weaken or break down the linkages between tannins and cell wall. In this way, the partial or overall degradation of material cells allows leasing tannins from plants. Cellulase enzyme usually exhibits high efficiency in tannin extraction from seaweed at other temperatures compared to other enzymes. Enzyme-assisted extraction permits the extraction of tannins with high stability and high activity. This is fully feasible because catalyze reactions and regioselectivity of enzymes can take place in aqueous medium under mild conditions.

Disrupting the structural integrity of the material cells by enzymes is the easy method and low solvent consumption. However, a clear understanding of the catalytic property, the material structure, the mode of action and the enzyme conditions is necessary. The effectiveness of tannin extraction mainly depends on ionic strength, ions in solution and characters of proteins and tannins.

x) Gamma Assisted Extraction

Gamma-assisted extraction is a new method using Co-60 gamma radiation for tannin extraction. Co-60 gamma radiation increases the effectiveness of the tannins from *Anacardium occidentale*. For tannin extraction, materials can be directly irradiated by gamma radiation, materials are then soaked in specific solvent or the mixture of materials and solvents is directly irradiated by gamma radiation.

Opportunity and Challenge of tannin extraction

Tannins commonly exist in the nature and they are bioactive diversity such as antioxidant, antifungal, anti-virus, anti-inflammatory and anti-parasitic. Tannins are used in numerous fields, for example, feed of ruminants, monogastric animals, etc. Each method of tannin extraction has difficult and easy points for example, high investment, difficult operation, or popular equipment. However, all methods are based on the dynamic and static maceration. All modern techniques are supporting and improving traditional extraction methods. At the same time, the demand using tannins in the market is very huge. Therefore, opportunity of tannin extraction and commercializing tannins are great. However, large scale production of tannins greatly differs in comparison to small-scale production. The content, structure, and activity of tannins in plants depend on growth time of plants, habitat place of plants, species, the parts of the plants, etc. Hence, stabilizing the content, structure and the activity of tannin is difficult. Tannin content is decreased in storage time of materials. The results of the tannin activities are mainly at the laboratory level. At the same time, the modern method of tannin extraction needs high investment, good manpower and a great market potential.

Applications of Tannin

- The tannin has the potential to use for various purposes such as leather processing industry, wood industry, anti-corrosion of metal, water treatment plant, animal food, medical science and the production of advanced materials.
- Tannins are the major tanning material for converting animal hide/skin into leather.
- Efficient extraction of tannins might also help to reduce the use of synthetic toxic materials. Gallic acid used for manufacturing plastic resins and adhesives can be used to produce different types of composites. The cross-linking of tannin by formaldehyde in the reaction of polycondensation can provide exterior category wood adhesive that can be used as adhesive for

particle board which can replace a significant portion of synthetic phenol from phenol-formaldehyde resin.

- While soluble, astringent materials are found in some plants like tea and coffee, tannins are supplemented with various processed foods, including ice cream and caramel.
- Apart from tanning, tannins are also used in dyeing, photography, refining beer and wine as well as an astringent in medicines.
- Tannins may be employed medicinally in antidiarrheal, hemostatic and anti hemorrhoidal compounds. The anti-inflammatory effects of tannins help control all indications of gastritis, esophagitis, enteritis and irritating bowel disorders. Tannins can also be effective in protecting the kidneys. Tannins have been used for immediate relief of sore throats, diarrhea, dysentery, hemorrhaging, fatigue and skin ulcers.
- Hydrolyzable tannins can be used as a wood preservative and thus, increase the durability of wood by protecting against fungal attacks.
- Tannins are used as food preservatives, packaging materials and food enhancement which owe to their protective nature.
- Used as nano-packaging material which offers high density, enhanced surface hydrophobicity, improvement in air-barrier and antioxidant properties. Simultaneously, nanocellulose tannin based films are active packaging materials which also provide a green, sustainable, nontoxic packaging source for food and pharmaceutical products.

Suggested Readings

Alfredo Aires 2020 *Tannins - Structural Properties, Biological Properties and Current Knowledge,* London: IntechOpen

Andrzej Szczurek 2021 *Perspectives on Tannins* Switzerland: MDPI publisher of open access

Cheryl Anne Combs 2016 *Tannins: Biochemistry, Food Sources and Nutritional Properties,* New Yark: Nova Science Publishers

Harinder PS Makkar 2003 *Quantification of Tannins in Tree and Shrub Foliage - A Laboratory Manual,* Springer: Dordrecht

Procter HR 2013 *The Chemistry of Tannins and Their Derivatives - A Historical Article on Leather Production,* Thrissur: Read books

CHAPTER 5
Natural Dyes

V. Priyanka and B. Vinothini
Forest College and Research institute, Mettupalayam

Natural dyes are derived from naturally occurring sources such as plants, insects, animals and minerals without any chemical treatment. A spectrum of beautiful natural colours ranging from yellow to black exists in the plant and animals. They are known for their use in colouring of food substrate, leather as well as natural protein fibres like wool, silk and cotton as major areas of application since pre-historic times as it was practiced during the Bronze Age in Europe. Recently, interest in the use of natural dyes has been growing rapidly due to the result of stringent environmental standards imposed by many countries in response to toxic and allergic reactions associated with synthetic dyes. Research has shown that synthetic dyes are suspected to release harmful chemicals that are allergic, carcinogenic and detrimental to human health. On the other hand, natural dyes are non-allergic, non- toxic and environment friendly; for example, turmeric, the brightest of naturally occurring natural dyes is a powerful antiseptic which revitalizes the skin, while indigo gives a cooling sensation. They produce very uncommon, soothing and soft shades unlike synthetic dyes. After a century of neglect, there is a plenty of scope for rapid developments in natural dyes.

India is a country of culture and colour. Since antiquity, festivals in India are associated with colour and these were sacredly extracted from plants and animals. A systematic, but secretive art form limited the shades to a few. This is one of the reasons for limited use. If these major gaps are answered, natural dyes extraction and application can be taken at commercial level and can enter into global market. India has a rich biodiversity and there is ample scope to explore and revive the application of natural dyes from plant kingdom. Trees are also one of the important sources of natural dyes. As many as 10,000 different shades have been developed from flowers, seeds and barks of plant sources like *Butea monosperma*, *Bixa orellana*, *Tagetus erecta*, *Indigofera tinctoria*, *Eclipta prostata*, *Eucalyptus* spp., *Terminalia* spp. and *Caesalpinia sappan*. Thus, many natural resources which contain many useful dye pigments are presently being wasted. Trees are also a source of alternate for vegetable dyes, the use of which questions the food security of our country. Several commercial natural dyes are obtained from trees, like red dye from Brazil wood (*Caesalpinia echinata* Lam.), brown dye from catechu (*Acacia catechu* L.) and yellow dye from old fustic (*Maclura tinctoria*). However, commercialization of dyes can be successful with systematic and scientific approach for identification of resources, extraction, purification, chemical structure elucidation and thereby promoting the use of natural dyes.

Status of Natural Dyes in India

India has a very rich tradition of using natural dyes. The art and craft of producing natural dyed textile has been practiced since ages in many villages by traditional expert crafts persons in the country. Natural dyes, when used by them have many limitations of fastness and brilliancy of shade. However, when used along with metallic mordants they produce bright and fast colour. The use of metallic mordants is not always eco-friendly, but the pollution problems created by metallic mordants are of very low order and can be easily overcome. Therefore, instead of using unsustainable technology for producing colours one can use mild chemistry to achieve almost similar results.

Certain chemical compounds used in chemical dyes are found to be carcinogenic, mutagenic and sensitizing and as a result, they have been banned. Azo dyes have been red listed. Therefore, manufactures of handloom goods and those promoting their exports have come to realize that the developed countries like US, Germany, Austria and Denmark patronizing Indian handloom prefer garments coloured with natural dyes as compared to synthetic dyes. This itself provided the main thrust for the retrospection by the dyeing industry and the revival of interest in natural dyes and the introduction of eco-friendly natural dyes as an alternate to the existing synthetic dyes. There is a growing awareness and public concern in recent years over the issue like environmental pollution and hazards arising out of the textile dyeing industry. Dyeing of fabrics with natural dyes, a rich traditional craft is now therefore being revived.

Classification of Natural Dyes

Natural dyes can be classified in various ways, viz., Based on Colours, Chemical constitution, Application and Origin.

a) Classification based on colour

Natural dyes are frequently categorized on the basis of the colour that they impart to the substrate (Fig. 5.1).

Fig. 5.1 Classification of dye based on colour

i) Red

The Colour Index lists 32 red natural dyes. The prominent among them are Madder (*Rubia tinctorum*), Manjeet (*Rubia cordifolia*), Brazil wood/Sappan wood (*Caesalpina sappan*), Morinda (*Morinda citrifolia*), Cochineal (*Coccus cacti*) and Lac dye (*Coccus lacca*). All these dyes are based on anthraquinone molecule except Brazil and Sappan wood based dyes. These dyes are prone to oxidation and hence are not suitable. The strong and almost fadeless cotton dye known as Turkey Red was developed in India and spread from there to Turkey. It involved about twenty separate processes using wood, oil and rancid fat, charcoal, cow/sheep/dog dung and liquid content of animal stomach. Only the dyers and their families not surprisingly, occupied villages where the process was carried out. The use of madder declined abruptly following the development of synthetic alizarin in 1869.

ii) Blue

The Colour Index lists only 3 natural blue dyes; namely natural indigo, sulphonated natural indigo and the flowers of the Japanese Tsuyukusa. The only viable choice among the blue natural dyes is indigo. Natural indigo is obtained by fermenting the leaves of various species of Indigofera, running off the liquor and oxidizing it to precipitate the dye. Woad (*Isatis tinctoria*) is another source of indigo. The main ingredients of natural indigo are indigotin and indirubin. Natural indigo has higher affinity and the dyed fabrics have better fastness.

iii) Yellow

Yellow is the most common colour in the natural dyes. However, most of the yellow colourants are fugitive. The Colour Index lists 28 yellow dyes. Some of the important yellow dyes are obtained from Berberry (*Berberis aristata*), Tesu flowers (*Butea frondosa*, *Butea monosperma*) and kamala (*Mallotus philippensis*). Other sources of yellow dyes are Black oak (*Quercus velutina*), turmeric (*Curcuma longa*), weld (*Reseda luteola*) and Himalayan rhubarb (*Rheum emodi*).

iv) Green

Plants that yield green dyes are rare. Both woad (*Isatis tinctoria*) and indigo have been used since ancient times in combination with yellow dyes to produce green shades.

v) Black and Brown

Cutch is an ancient brown dye from the wood of acacia trees, particularly *Acacia catechu*. According to Colour Index, 6 dyes are available in black colour. Some common examples of black colour are roots of iris plant, lac, carbon and caramel.

vi) Orange

Dyes that create reds and yellows can also yield oranges. The sources for a natural orange dye are Berberry, Annatto, Sweet pepper, Blood roots, etc.

b) Classification based on chemical constitution

Natural organic dyes cover a wide range of chemical classes, *viz.* anthraquinone, indigoid, naphthoquinone, carotenoid, flavone, di-hydro-pyransantho cyanidin and flavonols.

i) Anthraquinone dyes

Few most important red dyes are based on the anthraquinone structure and are obtained both from plants as well as animals or insects. Examples of anthraquinone dyes are madder (extracted from the root, known as the alizarin), lac dye (animal dye), kermes (insect dye and the colouring matter is kermesic acid), cochineal (obtained from an insect and main colouring component is carminic acid).

ii) Indigoid dyes

Two very important natural dyes possess indigoid structure, namely Indigo and Tyrian purple. Indigo is perhaps the oldest natural dye used by man. The dye from the plant *Indigofera tinctoria* has been used pre-historically in India, where it probably originated. It is known in India since 4000 years as Naphthoquinone dyes. The most prominent member in this class is lawsone or henna. It is obtained from *Lawsonia inermis*, cultivated mainly in India and Egypt. Another similar dye is juglone obtained from the shells of unripe walnuts.

iii) Carotenoid dyes

The class name is derived from the orange pigment found in the carrots. In these, the colour is due to the presence of long conjugated double bond. Some common examples of the carotenoid structure based dyes are Annatto, Saffron, etc.

iv) Flavonoid dyes

Flavonoids, which yield yellow colour, can be classified under flavones, isoflavones, aurones and chalcones. Flavones are colourless organic compounds. Most of the natural yellows are derivatives of hydroxyl and methoxy substituted flavones and isoflavones. Common example is weld containing luteolin pigment.

v) Dihydropyran dyes

The substituted dihydropyrans, *viz.* Haematin and its leuco form, haematoxylin are closely related to the chemical structure of the flavones. These are important natural dyes for dark shades. Logwoods, brazil wood and sappan wood are the common examples. Anthocyanins are the most important flavonoid colourant in plants and mostly occur in floral tissues.

c) Classification based on application

Natural dyes are also classified based on the method of their application, *viz.* mordant dyes, direct dyes, vat dyes, acid dyes and basic dyes and disperse dye.

i) Mordant dyes

These essentially require a mordant for application because of their lack of affinity for the fibre. According to definition, these dyes form complexes with the mordant. Three types of mordants are used, *viz.* i) Metallic mordants such as metal salts of aluminum, copper, tin, etc., ii) Tannic mordant such as tannic acid, eg. myrobalan and sumach and iii) Oil mordant which forms complex with main metal mordant.

ii) Vat dyes

The vat dye derived their name from the wooden fermentation vessel called Vat, which was at one time used for reducing the dye in order to convert it into soluble form. This process of solubilization is called vatting and the soluble form of dye is called leuco dye. The soluble leuco form has affinity for natural fibres, which can be oxidized back on fibre on exposure to air. The vatting is also carried out by treatment with reducing agent, such as sodium hydro-sulphide and alkali, such as sodium hydroxide, eg. indigo

iii) Direct dyes

Direct dyes possess affinity for cellulosic fibres without any pretreatment to dye or to the fibre. Many natural dyes belong to this class. The most common examples are turmeric, harda, pomegranate rind and annatto.

iv) Acid dyes

These possess either sulphonic or carboxylic groups in structure and are most suitable for dyeing of wool and silk from acidic medium, eg. saffron.

v) Basic dyes

Basic dyes are cationic dyes which on ionization develop coloured cations and form electrovalent bond with the COOH groups of wool and silk, eg. berberine.

vi) Disperse dyes

A disperse dye has a low relative molecular mass, low solubility and no strong solubilizing groups. These dyes have hydroxyl and/ or amino groups which imparts some solubility to the dye molecule. One such dye could be lawsone. Many other flavones and anthraquinone dyes can be classified as disperse dyes.

d) Classification based on origin

There are primarily three sources from which natural dyes are extracted, *viz*. plants, minerals and animals.

i) Plants

Various parts of many plants are found potentially rich in natural dyes, plant parts like root, bark, stem, seeds and fruit can be used to extract colour. Some plants may possess more than one colour depending upon from which part of the plant they are extracted.

ii) Minerals

These colourants are derived from the natural mineral sources. Mineral colours are produced from purified natural organic compounds. Some of the important mineral colorants are chrome yellow, iron buff, napkin yellow and manganese brown.

iii) Animals

Animals are also rich source of natural dyes. Dyes normally are being extracted from the dried body of the insects; common examples are lac and cochineal.

Extraction Techniques

Natural Dyes cannot be used directly from their renewable sources. Safe and cheap extraction of main colouring component is most important without affecting the extraction conditions and avoiding any contamination in various extraction techniques. Several extraction methodologies for natural dye that comply with both consumer preference and regulatory control and that are cost effective are becoming more popular. The different extraction techniques are simple aqueous methods, enzymatic extraction, fermentation, complicated solvent systems, ultrasonic extraction in sonicator and supercritical fluid extraction.

i) Simple Aqueous Method

Natural dye sources are mixed with required amount of water and boiled for optimum time which is 60 min in most cases. The content is cooled to room temperature and filtered. The filtrate is used as a dye for dyeing (Fig. 5.2).

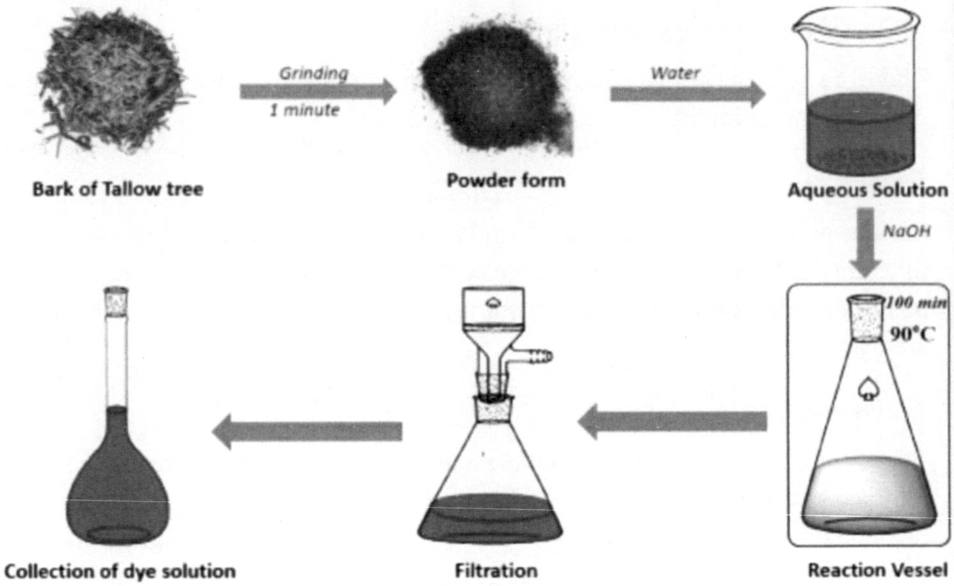

Fig. 5.2 Simple Aqueous Method of dye extraction

ii) Enzymatic Extraction

As plant tissues contain cellulose, starch and pectins as binding materials, commercially available enzymes including cellulase, amylase and pectinase have been used by some researchers to loosen the surrounding material leading to the extraction of dye molecules under milder conditions. This process may be beneficial in the extraction of dye from hard plant materials such as bark, roots, etc.

iii) Fermentation

This method of extraction uses the enzymes produced by the microorganisms present in the atmosphere or those present in the natural resources for assisting the extraction process. Indigo extraction is the most common example for this type of extraction. Freshly harvested indigo leaves and twigs are soaked in warm water (about 32° C). Fermentation sets in and the colourless indigo containing glucoside indican present in the leaves is broken down into glucose and indoxyl by the indimuslin enzyme present in the leaves.

Fermentation is complete in about 10-15 hours and the yellow liquor containing indoxyl is then taken to beating vats where indoxyl gets oxidized by air to the blue-coloured insoluble indigotin which settles down at the bottom. It is collected, washed and after removing excess water is pressed into cakes. Extraction of indigo from other indigo containing plants such as woad is also carried out by fermentation.

The fermentation method is similar to aqueous extraction with the exception that this method does not require high temperature. The microorganisms disintegrate the colouring matter binding substances in natural way. Long extraction time, need of immediate extraction of pigments after harvesting and foul smell due to microbial action are some of the disadvantages of this method.

iv) Complicated solvent systems

The dried material (leaves, roots, barks, wood, resinous secretion of insects, etc.) are ground to very fine particles. The crude dried powder is weighed and solvent is extracted using soxhlet apparatus, steam heated extractor. Different solvents (such as acetone, chloroform, ether, n-hexane, alcohol, soda ash, etc.) are used for extraction. The process is carried out for 4 h. The dye extract is evaporated in an evaporating dish over a water bath. After evaporating to dryness, the solute is weighed and the percentage yield is calculated, eg. extraction of lawsone from henna.

v) Ultrasonic extraction in sonicator

The combined effect associated with ultra sound energy is of cavitation's, compressions, rarefactions and micro-streaming results in intermolecular tearing and surface scrubbing. In particular, it has been noted that, some reactions when exposed to ultrasonic energy become faster with lower temperature that is the most beneficial effect as it reduces processing time and energy consumption and improves product quality, eg. Sonicator is used for sappan wood extraction in water using ultra-sound energy. To extract the dye faster and in a more efficient manner, sonicator with ultra sound energy of 20 kHz frequency is used.

vi) Supercritical fluid extraction

One of the major technologies that have emerged over the last two decades as an alternative to the traditional solvent extraction of natural products is the supercritical fluid extraction technique. It uses a clean, safe, inexpensive, nonflammable, nontoxic, environmental friendly, non-polluting solvent such as CO_2. Besides, the energy cost associated with this technique is lower than the costs for traditional solvent extraction method. Supercritical fluid extraction technology is thus increasingly gaining importance over the conventional techniques for extraction of natural products. CO_2 is the most widely investigated supercritical fluid extraction solvent since its critical temperature (Tc = 31.1°C) makes it an ideal solvent for extracting materials that are thermally liable. CO_2 is an ideal solvent in the food, dye, pharmaceutical and cosmetic industries, where it is essential to obtain final products of a high degree of purity eg. supercritical CO_2 extraction of lycopene and β-carotene from ripe tomatoes.

Techniques for Standardization of Natural Dyes

Standardization of natural dye should be done by conversion of natural dye extracts into a fine powder, paste or solution form. Only water as such or at different pH and temperature would be used for colouring matter from the raw materials. Parameters such as pH, time and temperature of extraction should be worked out for optimization to achieve standardization. Standardized natural dye should be pure and does not change appreciably due to atmospheric conditions. For the determination of standardized natural dyes accurately, it is necessary to compare it with some known standard of purity. Standardization includes determination of extractive values using various solvent.

- **Ash values**: Determination of ash value is useful for determining low grade extracts.
- **Rf values**: Paper chromatography and thin layer chromatography is widely used as a mean of assessing quality and purity. Rf values of natural dyes are determined under specific conditions and can be used for standardization by an aid to identification.
- Most of natural dyes are pH sensitive and they change colour with changing pH, use of buffer solution for extraction of the dye is first step towards standardization.
- Same natural colour extracts will not give the original colour of extract on the fabric. Therefore, reproducibility of natural dye shade is very important in the process of standardization.
- Parameters such as pH, temperature of extraction and conditions for getting the maximum colouring matters should be standardized.
- Considerable standardization with respect to moisture, dye and ash content, particle size, pH of dye, storage stability should be done.

Standardization of Dye Extraction

The natural dye source (leaves, roots, wood, etc) is collected, dried in shade and powdered sample is boiled in various solution medium (aqueous, alkaline, acid, etc) in order to optimize the method of extraction. The optimizations are based on the optical density of the liquid, before and after dyeing using spectrophotometer and visual appearance.

- **Optimization of alkali / acid concentration:** Optimum alkali/ acid concentration for 100 ml of water, for a pH to be standardized.
- **Optimization of extraction time and dyeing time:** Boiling for longer duration increases dye concentration. High dye yield is found after boiling for optimized time. High absorption is found for optimized dyeing time, above which dye depletion is observed.
- **Optimization of concentration of mordants:** The concentration of mordants is optimized by considering the visual appearance and dye absorption.
- **Optimization of mordanting methods:** Sometimes more than the mordant, the mordanting methods contributes towards the production of darker or lighter and faster shades.

Applications of natural dyes

The development of natural pigments application in wellness industries especially food, cosmetics and pharmaceuticals has been established attributed by their nature abilities in various aspects.

i) Food industry

The necessity of the use of colourants in food substances has been highlighted to maintain original food appearance, assure colour uniformity, intensify foods normal colour, protect characteristics of food such as flavour and vitamins and increase food acceptability. Colourants from natural sources as food additive are able to prevent food spoilage caused by fungi, viruses and bacteria since most of them, including anthocyanins and carotene, possess antagonistic activity. The examples of natural colourants in foods are annatto pigment in cheddar cheese and marigold flower extract in poultry products.

ii) Cosmetic industry

Natural dyes are commonly applied in cosmetic industry due to their privileges of no side effects, compatible with all skin types, UV protection, not tested on animals and anti-aging properties. Numerous cosmetics in skincare, hair care, fragrance, etc. has been developed. Yellow pigment of turmeric is a powerful antiseptic for skin revitalization and antioxidant ability of carotenoid from annatto oil provide benefits as body care products besides give sunny colour to creams, shampoos and lotions. Roselle colour is often used as natural colourants in lipstick due to the effectiveness of anthocyanin content in minimizing wrinkles, fine lines and dryness of lips. The orange colour from ß-carotene of carrot oil can metabolize into vitamin A which plays an important role in anti-aging, revitalizing and rejuvenating agent.

iii) Pharmaceutical industry

Generally in pharmaceutical industry, the formulation of tablets, tablet coatings, capsules, liquid orals, tooth paste and ointment are coloured to increase the aesthetic appearance, prolong the stability, produce standard preparations or for identification purposes. The active constituents in natural pigments are utilized to meet the demand in medicine.

For example, carotenoids are used as vitamin supplements, bixin is applied as one of the natural colouring agents for ointments and plasters and film coating for tablets employed chlorophyll, carotenoids and anthocyanins as basic ingredients.

iv) Textile industry

Natural dyes are having wide application in the colouration of most of the natural fibres, e.g. cotton, linen, wool and silk fibre and to some extent for nylon and polyester synthetic fibre. However, the major issues for natural dyed textiles are reproducibility of shade, non-availability of well-defined standard procedure for application and poor lasting performance of shade under water and light exposure. The commonly used natural dyes are haldi, babul, madder, pomegranate rind, marigold and sappan wood.

Advantages of natural dyes

- It is not harmful to health
- It has no effluent generation
- It has mild dyeing condition
- It is a renewable source
- No allergic, carcinogenic or toxic reactions
- It has high UV ray absorption capacity

Drawbacks of natural dyes

Natural dyes are considered to be an eco-friendly alternative for dyeing of textile materials, especially natural fibre textiles. However, there are many limitations in the usage of natural dyes which are listed below.

i) Tedious application process

Natural dyes require a longer dyeing time in comparison with synthetic dyes as very often an additional mordanting step is required. Use of raw dye-bearing materials ensures authenticity but at the same time involves additional dye extraction steps that require time and separate set-up. Natural dyes in this form are also not suitable for use in many commercial textile dyeing machines which makes the process labour intensive. In an industrial set-up, disposal of solid residual biomass is also problematic. Purified extracts, although suitable for machine application, are costly and not economical. Logistics for making agricultural by-products such as pomegranate rind, onion skins, or fruits and leaves of trees from the forests available for dyeing purposes are not in place which would have helped in reducing the costs. Exhaustion of most of the natural dyes on textile materials is poor in spite of using the mordants which leaves a large quantity in the dye bath after dyeing. Traditional dyers reused the dye bath but the colour obtained is different from the earlier lot which may not be acceptable by today's standards. All these increase the cost of naturally dyed textiles.

ii) Limited shade range

Shade range of natural dyes is limited. Out of the three primary colours-viz., red, yellow and blue, although there are several sources for red and yellow dyes, there is only one major source of the blue dye, indigo. As natural dyes differ in their application process, only few dyes can be applied in mixtures and differences in fastness properties further limit the choice. Even a common secondary colour such as green needs to be produced by overdyeing as blue dye indigo being a vat dye has an entirely different application process that increases the dyeing time and cost.

iii) Non reproducible shades

Difficulty in reproducing the shades is another major drawback of natural dyes which is caused by the inherent variations in the proportion of chemical constituents in the natural material and thus in its crude extract depending upon the maturity, variety and agroclimatic variations such as soil type, region and so on. Therefore, it is not possible to produce the same shade with a particular natural dye in every dyeing operation. Production of standardized dye powders is an expensive and complicated exercise as most of the natural dyes contain various chemical constituents. Some of the natural dyes are pH sensitive and tend to change the colour due to change in pH. As natural dyes tend to form coloured complexes with metal ions, the mineral composition of water may also cause shade variations. Hence even the same standardized powder may give different shades at two different places due to the differences in mineral content and pH of the water which makes it very difficult to reproduce the shades.

iv) Fastness properties

Colour fastness properties of natural dyes are a cause of concern. Only a few natural dyes possess fastness properties conforming to modern textile requirements. Restrictions on the use of certain metal salts for mordanting such as chromium, copper, tin, etc. by eco-standards has not only reduce the colour gamut of natural dyes but also create the task of producing shades with good fastness properties difficult. Improper application procedures used by certain practitioners of natural dyes are also sometimes responsible for poor fastness properties. Improvements and optimization of mordanting and dyeing procedures can help in solving this issue. Exploring the new sources for dyes can increase the number of dyes with better colour fastness properties.

v) Safety issues

Exploration of new sources for dyes can certainly help in increasing the shade range of natural dyes with good fastness properties. However, extensive research on the safety of these materials to humans and the environment would be needed before propagating their usage as everything of natural origin may not be safe. Use of metallic mordants also requires caution so as not to cause adverse health effects during handling. Precautions should also be taken to prevent pollution problems in the usage of these mordants and it should be ensured that the amount of restricted mordants in the dyed textile is within the limits set by eco-regulations.

vi) Characterization and certification issues

Although dyeing of textile fabrics with dyes obtained from various natural resources has been extensively investigated, little information is available on the identification and characterization of the natural dyes. Natural dyes, being plant metabolites, are present only in small amounts in dye-bearing materials along with large quantities of other non-dye materials. The dye content may vary according to the age, part of the plant and agroclimatic conditions and it is important to know the dye content in order to get reproducible shades. While procuring the dye materials, pricing should match the dye content and when powdered dye materials or extracts are used. It should be authentic. Thus, determination of dye content as well as characterization of dye material is important in the case of natural dyes. Absorption spectroscopy is very successfully used for measuring the dye content of synthetic dyes but has limited applicability for natural dyes as these dyes are usually not a single chemical entity but a mixture of closely related compounds and in many cases, there are no clearly defined absorption maxima. Therefore, development of process protocols for quick identification and characterization of natural dyes is very important for the sustainability of true natural dyed textiles. TLC and HPTLC techniques can be easily employed for quick identification of natural dyes by comparing with chromatographic fingerprints of authentic natural dye samples. Establishment of some certification system such as those followed for organic crop products would be very helpful in this regard as that can ensure both authenticity and quality of natural dyes.

Suggested Readings

Daniel M Bhattacharya SD Arun Arya and Vinay M Raole 2021 *Natural dyes: scope and challenges*, Jodhpur: Scientific Publishers

Farooqi MIH 2008 *Dictionary of Indian Plant Gums, Resins, Dyes and Related Products,* Lucknow: Sidrah Publishers

James P Bernard 2017 A blog on Brief History of Natural Dyes

Padma S Vankar 2016 *Handbook on Natural Dyes for Industrial Applications (Extraction of Dyestuff from Flowers, Leaves, Vegetables),* New Delhi: NIIR project consultancy services

Rebecca Desnos 2020 *Botanical Dyes on Wood,* Rebecca Desnos

Sasha Duerr 2011 *The Handbook of Natural Plant Dyes,* Portland: Timber Press

CHAPTER 6
Essential Oils

M.P. Divya and P. Hemalatha
Tamil Nadu Agricultural University, Coimbatore

Essential oils (EOs) are basically phytoconstituents. They are produced by steaming or pressing different parts of a plant (flowers, bark, leaves, roots and fruits) to extract the fragrant components. Cosmetics, pharmaceuticals, food, drinks, cleaning products and the perfume business all employ essential oils. There are about 400 essential oils in use. Essential oils (EOs) are aromatic, partially water-soluble, inflammable, volatile and soluble in alcohol liquids obtained from different plant materials by different suitable methods. Essential oils can vary according to temperature, soil conditions, altitude and the country of origin. It has taste and odour because they contain volatile aroma-active compounds (i.e., molecules that elicit a distinctive taste and smell). Most EOs contains hydrocarbons and oxygenated compounds, mainly as terpenes and terpenoids. The majority of essential oils include antibacterial, antioxidant and functional properties Despite their enormous potential as preservatives and functional components, EOs is limited in their industrial application due to their high volatility, hydrophobicity, oxidation susceptibility, low stability and solubility.

The important essential oil yielding plants along with parts used, method of extraction and the presence of chemical compounds are given in Table 6.1.

Table 6.1 Sources of Essential Oils

Plant species	Parts to be used	Methods of extraction	Chemical compounds
Centella asiatica (Centella)	Leaves	Cold Press Method	Sesquiterpene hydrocarbons, β-Caryophyllene, α-Humulene, Trans-β-Farnesene, Germacrene-D and Caryophyllene oxide
Myrrhis odorata (Fennel)	Leaves	Hydrodistillation	E-Anethole, Methyleugenol and E-Nerolidol, Germacrene-D
Petroselinum crispum (Parsley)	Seeds	Hydrodistillation	1, 3, 8-p-Menthatriene, β-Phellandrene, Apiol, Myristicin and Terpinolene

Plant species	Parts to be used	Methods of extraction	Chemical compounds
Chrysothamnus nauseosus (Rabbit brush)	Flowers	Hydrodistillation	Apigenin-4′-Methyl ether, 3-O-acetyl-Taxifolin, Kaempferol-3-Methylether, 5-Dimethyl ether, β-Naphthol,7-Methyl-β-Naphthol
Ageratum conyzoides (Goat weed)	Flowers, Stem & Leaves	Hydrodistillation	Ageratochromene (Precocene II), 6-Demethoxy agerato chromene (Precocene I), Hexanal, α-Pinene, Camphene, β-Pinene, Myrcene, δ-2-Carene, α-Phellandrene, 2-Methoxy-4-Vinylphenol and β-Caryophyllene.
Thujopsis dolabrata (Hiba)	Wood	Steam Distillation	1-Ethyl-3-Methylimidazolium Methyl phosphonate, Thujopsene, Cedrol and Hinokitiol.
Cryptomeria japonica (Japanese Cedar)	Leaves	Water Distillation	α-Pinene, Car-3-ene and Sabinene, α-Terpinene.
Cupressus macrocarpa (Monterey cypress)	Leaves	Water Distillation	α-Pinene, and β-Pinene Terpinen-4-ol, α-Phellandrene, α-Citronellol, and Citronellal
Lavandula angustifolia (Lavender)	Flower	Steam Distillation	Linalool, Camphor, Terpinen-4-ol, Linalyl Acetate, β-Ocimene and 1,8-Cineole, α-Thujene, β-Pinene and Limonene
Ocimum basilicum (Basil)	Leaves & Flowers	Hydrodistillation, Steam Distillation & Solvent Extraction	Methyl Eugenol (78.02%), α-Cubebene (6.17%), Nerol (0.83%) and ε-Muurolene (0.74%)
Mentha piperita (Mint)	Leaves	Hydrodistillation & Steam Distillation	Menthol (40.7%) and Menthone (23.4%)
Rosemary officinalis (Rosemary)	Leaves	Hydrodistillation & Steam Distillation.	1, 8-Cineol, Camphor, α-Pinene, Limonene, Camphene and Linalool.
Salvia officinalis (Sage)	Leaves	Hydrodistillation	Cis-thujone, Camphor, Cineole, Humulene, Trans-Thujone, Camphene, Pinene, Limonene, Bornyl Acetate and Linalool

Plant species	Parts to be used	Methods of extraction	Chemical compounds
Thymus vulgaris (Thyme)	Leaves, Stem & Flowers	Hydrodistillation & Steam Distillation.	P-Cymene, γ-Terpinene and Thymol
Laurus nobilis (Bay leaf)	Leaves	Steam Distillation	Eucalyptol, α-Terpinenyl Acetate, Terpinene-4-ol and Sabinene
Sassafras albidum (Sassafras)	Root	Steam Distillation	Safrole, Camphor and Methyleugenol
Cinnamomum verum (Cinnamon)	Tender Twigs, Pedicels of Buds, Flowers & Fruits	Steam Distillation	Cinnamaldehyde, Camphor, Cinnamyl-Acetate, Caryophyllene, Trans α-Bergamotene, Caryophillene oxide, Linalool, Geraniol, Bornyl Acetate, α-Cubebene, γ-Elemene, α-Copaene, Guaiol and Eugenol
Backhousia citriodora (Lemon myrtle)	Leaves	Hydro Distillation	Linalool, Citronellal, E-Isocitral, Geraniol, Ethyl Geranate, Caryophyllene and Spathulenol.
Eucalyptus spp.	Leaves	Water Distillation & Steam Distillation	Cineole, Methyl Amyl Acetate, Aromadendrene, and Viridiflorol
Leptospermum scoparium	Leaves	Water Distillation	Calamenene, Ledene, α-Selinene, and α-Copaene. Leptospermone, Isoleptospermone, Flavesone and γ-Muurolene.
Pimenta dioica	Leaves	Water Distillation	Eugenol, α-Pinene, Caryophyllene, Limonene and 1,8-Cineole
Melaleuca alternifolia (Tea tree)	Leaves	Steam Distillation	Terpinen-4-ol Chemotype, a Terpinolene Chemotype, and four 1,8-Cineole Chemotypes
Abies alba	Leaves, Twigs, Cone & Needle	Steam Distillation	α-Pinene, Limonene and β-Pinene
Cedrus spp. (Cedar)	Leaves, Twigs, Cone & Needle	Steam Distillation	β-Himachalene, α-Himachalene and γ-Himachalene

Plant species	Parts to be used	Methods of extraction	Chemical compounds
Pinus spp. (Pine oil)	Leaves, Twigs, Cone & Needle	Steam Distillation	Turpentine and Terpenoids
Santalum album (Sandalwood)	Wood	Hydrodistillation & Solvent Extraction	α-Santalol and β–Santalol, Esters, Alcohols, Aldehydes
Juniperus communis (common juniper)	Leaves, Twigs, Cone & Branch	Hydrodistillation & Solvent Extraction	Tricyclene, Camphene, Sabinene, β-Pinene, Linaleol and Camphor
Casearia sylvestris (wild sage)	Leaves	Water Distillation	Caryophyllene, α-Humulene, Germacrene D, Bicyclogermacrene, Spathulenol, Caryophyllene oxide and Humulene epoxide
Casearia decandra (wild honey tree)	Leaves	Water Distillation	Monoterpenes, Oxygenated Monoterpenes, Sesquiterpenes, Oxygenated Sesquiterpenes, Phenylpropanoids and Ketone
Ocotea catharinensis (black cinnamon)	Leaves	Water Distillation	α-Pinene, β-Pinene, β-Caryophyllene & Germacrene-D.
Nectandra grandiflora	Leaves	Water Distillation	Sesquiterpenoid and Dehydrofukinone
Ocotea odorifera	Leaves	Water Distillation	Safrole, γ-Cadinene, Camphor and α-Copaene
Drymis angustifolia	Leaves	Water Distillation	Bicyclogermacrene, Sabinene and Myrcene, while DB oil was characterized by Cyclocolorenone, followed by Terpinen-4- ol and Alpha-gurjunene.

Extraction Methods

There are three types of distillation viz., Hydro distillation, Hydro and steam distillation & Direct steam distillation. These terms were first coined by Von Rechenberg, then it has become widely used in the essential oil market. All three approaches are subject to the same theoretical concepts that apply to two-phase system of distillation. The fundamental distinction is based on the materials are handled. Because some volatile oils cannot be distilled without decomposition, they must be obtained through expression or other mechanical processes. The majority of modern essential oil production in the perfume industry is done through extraction with volatile solvents like petroleum ether and hexane. The fundamental benefit of extraction over distillation is that a constant temperature (typically 50°C) can be maintained throughout the process. As a result, extracted oils have a more natural odour than distilled oils, which may have been chemically altered by the high temperature. This type of feature has high market value for the perfume, yet the conventional distillation method is less expensive than the extraction process.

Distilling volatile oil in the absence of air is known as destructive distillation. When Pinaceae or Cupressaceae wood or resin is heated without oxygen, breakdown occurs and a number of volatile chemicals are released. Depending on the wood used, the condensed volatile matter separates into two layers: an aqueous layer containing wood naptha (methyl alcohol) and pyroligneous acid (crude acetic) and a tarry liquid containing pine tar, juniper tar or other tars. Dry distillation is commonly done in retorts and if the wood is chipped or coarsely powdered and the heat is applied quickly enough, the yield can be as high as 10% of the wood weight utilized.

a) Hydro Distillation

The material is totally immersed in water, which is then boiled using direct heat, a steam jacket, a closed steam jacket, a closed steam coil, or an open steam coil. The fundamental feature of this method is that boiling water comes into direct touch with the plant material. To prevent the charge from overheating while the still is heated by direct fire, proper precautions must be taken. When a steam jacket or closed steam coil is utilized, the risk of overheating is reduced; with open steam coils, this risk is eliminated. When using open steam, however, caution must be exercised to avoid the accumulation of condensed water within the still. As a result, the still must be thoroughly insulated. As the water boils, the plant material in the still must be stirred or agglomerations of dense material will form on the bottom and get thermally damaged. Certain mucilage-rich plant materials such as cinnamon bark must be powdered so that the charge can easily disperse in the water; when the temperature of the water rises, the mucilage from the ground cinnamon will be leached. The water-charge mixture's viscosity is substantially increased as a result, allowing it to burn. As a result, before doing any field distillation, a small-scale water distillation in glassware should be carried out to see if any changes occur during the distillation process. The Clevenger system is the laboratory device recommended for trial distillations. During water distillation, boiling water must keep all elements of the plant charge moving; this is feasible when the distillation material is loosely charged and maintains flexible in the boiling. Water distillation has only one distinguishing advantage: it allows the processing of finely powdered material or plant parts that, if exposed to live steam, would normally form lumps through which the steam could not pass. Other advantages of water distillation include the stills' low cost, ease of construction and suitability for field use. In several nations, these are still commonly utilized with portable equipment.

The main disadvantage of hydrodistillation is that it does not allow for total extraction. Furthermore, some esters are partially hydrolyzed and sensitive compounds such as aldehydes have a tendency to polymerize. Distilling water necessitates a larger number of stills, more room and more fuel. It necessitates a great deal of practice and knowledge with the method. The oil ingredients, which are high-boiling and slightly water-soluble, cannot be totally vaporised or require a substantial amount of steam. As a result, the procedure becomes unprofitable. For these reasons, water distillation is only employed when the plant material cannot be treated by water and steam distillation or straight steam distillation due to its composition (Fig. 6.1).

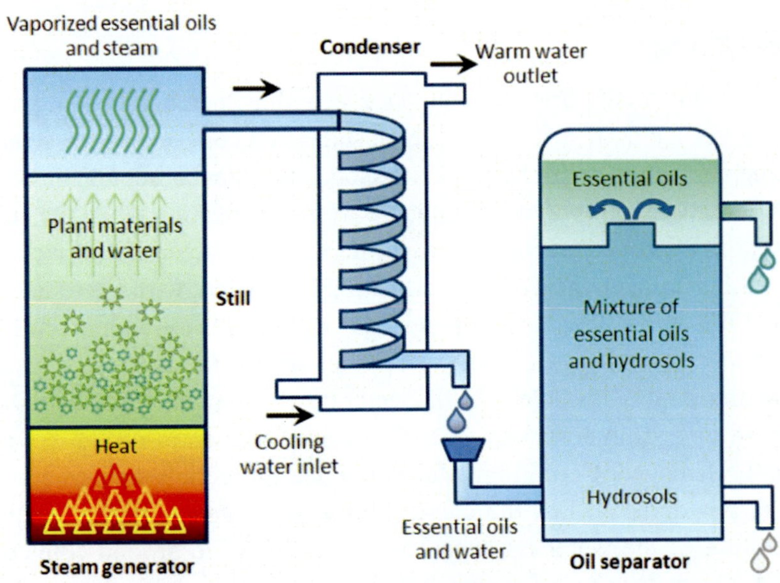

Fig. 6.1 Hydro Distillation

Mechanism of distillation

The primary physicochemical processes involved in hydro distillation of plant material are as follows: i) Hydro diffusion ii) Hydrolysis and iii) Heat-induced decomposition.

i) Hydro diffusion

Hydro diffusion is the process of essential oils and hot water diffusing across plant membranes. The steam in steam distillation does not enter the dry cell membranes. As a result, dry plant material can only be exhausted with dry steam after all the volatile oil has been released from the oil-bearing cells by thorough comminution of the plant materials. When the plant material is wet with water, however, vapours are exchanged within the tissue based on their permeability while swelling. Plant cell membranes are nearly impervious to volatile oils. As a result, in the actual process, a portion of the volatile oil dissolves in the water present within the glands at a temperature of boiling water and this oil-water solution permeates the swollen membranes by osmosis, eventually reaching the outer surface, where the oil is vaporized by passing steam. Another feature of hydro diffusion is that the rate of oil vaporization is determined by the degree of solubility of the oil components in water rather than their volatility. As a result, the high-boiling, water-soluble parts of oil in plant tissue distil first, followed by the low-boiling, less water-soluble constituents. Distillation of un-comminuted fluid takes longer than comminute substance due to slow hydro diffusion rates.

ii) Hydrolysis

Hydrolysis is described as a chemical interaction between water and some essential oil ingredients. Esters are elements of essential oils that prefer to react with, especially at high temperatures, to generate acids and alcohols. As a result, if there is a lot of water, there will be a lot of alcohol and acid, resulting in a lower yield of essential oil. Furthermore, because this is a time-dependent reaction, the extent to which hydrolysis occurs is determined by the moment the oil and water come into contact. This is one of drawbacks of water distillation.

iii) Impact of Heat

At high temperatures, almost all of the components in essential oils become unstable. Distillation at low temperatures is required to obtain the highest-grade oil. The operating pressure in steam distillation determines the temperature, whereas the operating pressure in water distillation and water and steam distillation is normally ambient. All three phenomena, hydro diffusion, hydrolysis, and thermal decomposition occur at low temperatures by utilizing as little water as feasible in the case of steam distillation and fully comminuting and packing the plant material uniformly before distillation.

b) Water and Steam distillation

Steam can be created in a satellite boiler or within the still in water and steam distillation, but it must be separated from the plant material. Water and steam distillation, like water distillation are extensively employed in rural regions. Furthermore, it does not necessitate a significant increase in capital cost over water distillation. The equipment is also identical to that used in water distillation, with the exception that the plant material is held above the boiling water on a perforated grid. It's not uncommon for people who start with water distillation to move on to water and steam distillation. As a result, after a few batches of oil have been produced via water distillation, rural distillers recognize that the quality of the oil is poor due to its still notes (mild scent). As a result, several changes have been made. A perforated grid or plate is fashioned using the same still to lift the plant material above the water. The still's capacity is reduced as a result, but the oil quality is improved. If the amount of water available is insufficient to complete the distillation, a cohobation tube is attached and condensate water is manually supplied back to the still, guaranteeing that the water used as the steam source never runs out. Because the re-used condensate water will allow it to become saturated with dissolved constituents, more oil will dissolve in it, it is also anticipated that this will, to some extent, reduce the loss of dissolved oxygenated elements in the condensate water. (Fig. 6.2)

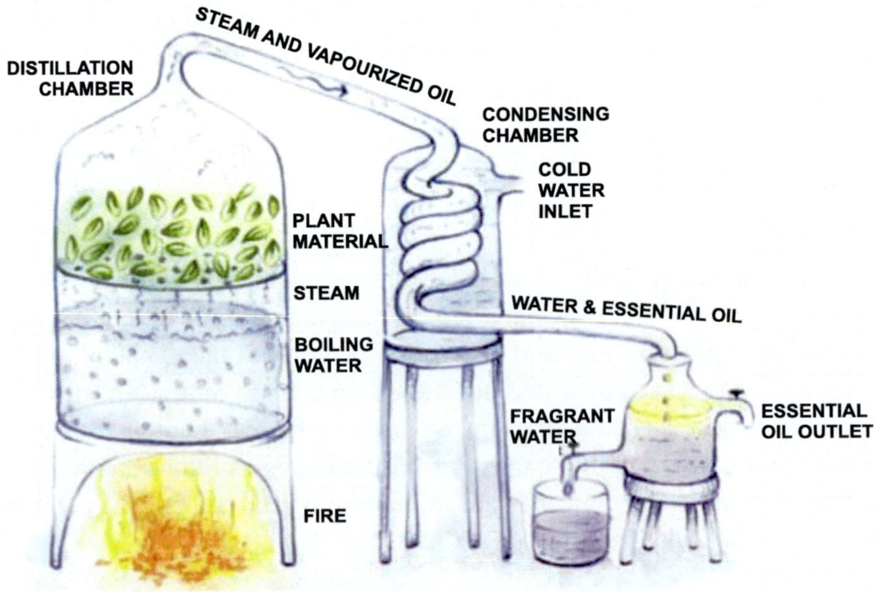

Fig. 6.2 Water and Steam distillation

Advantages of Water and Steam Distillation over Water Distillation

- Increased oil yield
- The volatile oil's components are less resistant to hydrolysis and polymerization
- If refluxing is kept under control, the loss of polar compounds is kept to a minimum
- The oil quality obtained through steam and water distillation is more consistent
- Steam and water distillation is more energy efficient than water distillation because it is faster. Many oils are now manufactured via steam and water distillation, such as lemongrass, which is produced in Bhutan using a rural steam and water distillation method.

Disadvantages of Water and Steam Distillation

Because of the low pressure of rising steam, high-boiling-range oils require more steam for vaporization, resulting in prolonged distillation processes. The wet plant material slows distillation because the steam must evaporate the water in order for these to condense further up the still. A baffle is used to prevent the lower plant material resting on the grid from being saturated by preventing the water from boiling too quickly and coming into direct contact with the plant material.

c) Direct Steam Distillation

Direct steam distillation is the process of distilling plant material with steam generated outside the still in a satellite steam generator, sometimes known as a boiler. The plant material is supported on a perforated grid above the steam input, just like in water and steam distillation. The amount of steam produced through satellite steam generation may be easily adjusted, which is a significant advantage. Because steam is created in a satellite boiler, the plant material is heated to not more than 100° C and as a result, thermal deterioration should not occur. The most widely established method for producing essential oils on a large scale is steam distillation. It is normal procedure in the flavour and fragrance supply industry.

The substantially larger capital cost required to build a steam distillation unit is an evident disadvantage. In some cases, such as large-scale production of low-cost oils (eg., rosemary, chinese cedarwood, lemongrass, spike lavender, eucalyptus, citronella, corn mint), the world market prices of the oils are barely high enough to justify their production by steam distillation without amortizing the capital expenditure required to build the facility over a 10-year or longer period (Fig. 6.3).

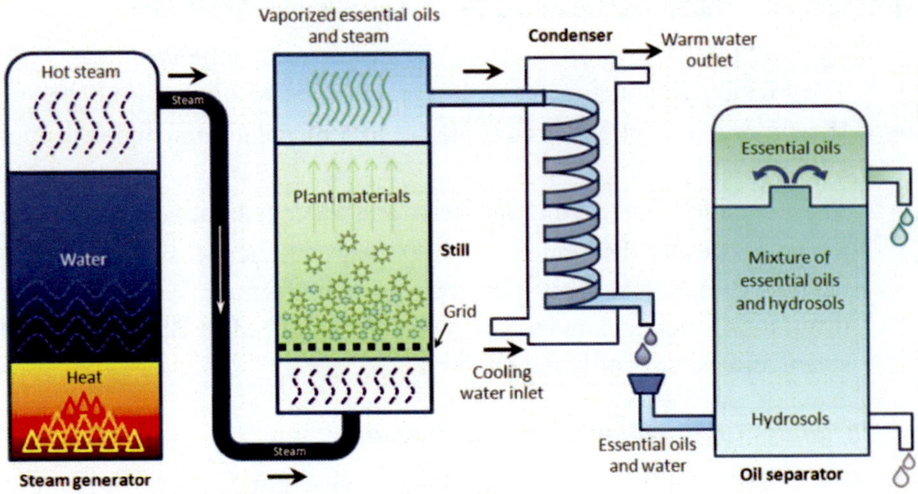

Fig. 6.3 Direct Steam Distillation

Advantages of Direct Steam Distillation

- The amount of steam can be easily adjusted
- Oil components do not decompose thermally
- It is the most widely accepted method for large-scale oil production and it outperforms the other two.

Disadvantage of Direct Steam Distillation

- Much higher capital expenditure needed to establish this activity than for the other two processes.

Other methods of essential oil extraction

i) Hydrolytic Maceration Distillation

Because their volatile components are glycosidically bound, certain plant materials require maceration in warm water before releasing their essential oils. For example, wintergreen (*Gaultheria procumbens*) leaves contain the precursor gaultherin as well as the enzyme peroxidase, which acts on the gaultherin to liberate free methyl salicylate and primrose when macerated in warm water.

ii) Essential Oil Extraction by Expression

Only citrus oils are produced by expression or cold pressing as it is commonly known. Any physical technique in which the essential oil glands in the peel are crushed or damaged to release the oil is referred to expression. One method used many years ago, especially in Sicily (spugna method), began with halving the citrus fruit and then removing the pulp with a sharpened spoon-knife. The oil is extracted from the peel by pressing it against a hard object of baked clay (concolina) put beneath a big natural sponge, or by bending it into the sponge. Oil made by this method has a stronger fruit odour than oil made any other way. Equaling (or the scodella method) is a second method that uses a shallow bowl of copper (or occasionally brass) with a hollow central tube; the equaling tool looks like a shallow funnel. The bowl has brass points with blunt ends on which the entire citrus fruit is rolled by hand until all of the oil glands have burst. Manual pressing is obviously impracticable because it is a very slow operation; for example, a single person may generate just 2-4 pounds oil per day using one of these hand processes. As a result, many machines have been developed throughout the years to either crush the peel of a citrus fruit or to crush the whole fruit and then separate the oil from the juice.

iii) Pelatrice Methodology

Citrus fruits are put into the machine's abrasive shell from a hopper in the pelatrice process. A slow-moving Archimedean screw rotates the fruits against the abrasive shell, rasping the fruit surfaces and causing some of the vital oil cavities on the peel to break and release their oil-water emulsion. This screw then delivers the fruit to a hopper, where abrasive spike-covered rollers burst the residual oil holes. A thin spray of water washes away the oil and water emulsion from the fruit. The emulsion then runs into a separator, which removes any particulates, before passing through two centrifugal separators in series to obtain pure oil.

iv) Sfumatrice Process

A copper chain is dragged by two horizontal ribbed rollers in the sfumatrice machinery. The peels are fed through these rollers, where they are squeezed and bent in order to extract their oil. The oil is rinsed away from the sfumatrice rollers with fine water sprays, just like in pelatrice. Again, the oil is processed through a separator before being fed through two centrifuges in series to produce purified oil. Sfumatrice was once the most popular method for isolating citrus oils in Italy; however, the pelatrice method now looks to be more popular.

v) Cold Fat Extraction of Essential Oils

Despite the introduction of the contemporary procedure of extraction with volatile solvents, the traditional method of enfleurage, which has been passed down from generation to generation and honed over centuries, continues to play an important role. Enfleurage is now only done on a wide scale in the Grasse region of France, with the exception of a few isolated cases in India where the method is still primitive.

Enfleurage follows a simple set of rules. Even after being harvested, certain flowers (Such as tuberose and jasmine) continue to develop and release perfume. Every jasmine and tuberose bloom are, in a sense, a miniature perfume factory that constantly emits tiny amounts of fragrance. Fat has a great absorption capacity and easily absorbs the aroma released when it comes into touch with fragrant flowers. Enfleurage is the use of this principle in a systematic way on a vast scale. Batches of newly harvested flowers are spread over the surface of a specially prepared fat base (Corps) for the entire harvest period, which lasts eight to ten weeks and left there before being replaced by fresh flowers.

The success of enfleurage is largely determined on the quality of the fat base used. Care is required when it comes to assembling the corp. It must be completely odourless and have the right consistency. If the corps is too hard, the blooms will not have enough contact with the fat, reducing its absorption potential and leading in a lesser flower oil output. If it is excessively soft, however, it will tend to swallow the flowers, causing the fatigued ones to adhere; when removed, the flowers will retain the adhering fat, leading in significant shrinkage and loss of tissue. The enfleurage process takes place in cool basements and each manufacturer must prepare the corps according to the temperature in the cellars throughout the flower harvest months.

vi) Enfleurage and Defleurage

Chassis are installed in every enfleurage structure to contain the fat corps during the operation. A chassis is a rectangular wooden frame. At the start of the enfleurage procedure, the fat material is applied to both sides of the glass plate with a spatula. With a layer of fat on the upper and lower sides of each glass plate, the chassis form airtight compartments when stacked one on top of the other. Freshly gathered flowers arrive every morning during the harvest and after being cleaned of impurities such as leaves and stalks, they are distributed by hand on top of the fat layer of each glass plate. Blossoms that have been moistened by dew or rain should never be used, as any moisture may turn the corpus rotten. Depending on the variety of flowers, the chassis are piled up and placed in the cellars for 24 h or longer. The latter are in direct touch with one fat layer (bottom one), which functions as a direct solvent, while the other fat layer (beneath the glass plate of the chassis above) absorbs just the flowers' volatile perfume.

After 24 h, the flowers have lost most of their oil and begin to wither, emitting an unpleasant odour. They must then be removed from the material, which is still done by hand, despite all efforts to deploy labour-saving equipment. Defleurage (careful removal of the flower) is almost as crucial as enfleurage, so the people doing this task must be competent and skilled. When the chassis is lightly struck against the working surface, most of the tired flowers will fall from the fat layer on the chassis glass plate, but because every single bloom and every particle of the flower must be removed, tweezers are required for this delicate procedure.

The chassis are recharged with fresh flowers immediately after defleurage, that is, every 24 h. The chassis are turned over for this, and the fat layer, which constituted the top (ceiling) of the little chamber in the previous procedure, is now directly charged with flowers. The complete enfleurage process takes roughly 70 days for jasmine, with the spent flowers being plucked daily and the chassis being replenished with new ones. The fat on the chassis is scratched over with metal combs and tiny furrows are drawn to modify and enhance the surface of absorption at the start of and multiple times during the harvest. The fat is significantly saturated with flower oil towards the end of the harvest and has the usual scent. After that, the scented fat must be removed from the glass plates that separate the chassis. It is scraped off with a spatula and then carefully melted and bulked in closed containers for this purpose. Pomade (Pomade de jasmine, Pomade de tuberous, Pomade de violet, etc.) is the final product.

Every chassis is charged with only around 360 g fat corps on each side of the glass plate at the beginning of the harvest, for a total of 720 g per chassis. For the entire time of enfleurage, which lasts from 8 to 10 weeks, each kg of fat corps should be in touch with roughly 2.5 kg (Ideally 3.0 kg) of jasmine flowers. The quantities fluctuate slightly depending on the flower. Because of the different manipulations, the fat corps has lost around 10% of its weight at the end of enfleurage.

vii) Hot Maceration Process

The long enfleurage period is halved in this procedure by immersing petals in molten fat heated to 45°-60° C for 1 to 2 h, depending on the plant types. The fat is filtered and removed from the petals after each dip. The fat is removed from the waste flowers and water after 10 to 20 immersions. Through a process of extraction and concentration at lower pressure, absolute of maceration is obtained from fat-containing oil. It is mostly utilised for particularly delicate flowers, such as lily of the valley, whose physiological activities are swiftly lost after harvest.

Modern Methods of Extraction

New techniques have been developed as a result of technological advancements, which may not be widely used for commercial production of essential oils but are considered valuable in certain situations, such as the production of expensive essential oils in their natural state without altering their thermosensitive components or the extraction of essential oils for micro analysis. The some of the modern methods of extraction are given below.

a) Headspace Trapping Technique

Headspace techniques are based on Henry's law, which states that the ratio of a dissolved substance in aqueous solution to the substance in the vapour in a sealed glass vial is constant depending on temperature.

i) Static Headspace Technique: Static Headspace (SHS) extraction technique is the selective technique exploiting the volatility of some compounds with respect to the sample matrix and their capacity to move from the sample to the headspace of a sealed container from where they are collected.

ii) Vacuum Headspace Technique: It is a dynamic headspace method (DHS), simultaneous distillation and extraction at atmospheric pressure (SDE) and simultaneous distillation and extraction under reduced pressure (SDEV).

iii) Dynamic Headspace Technique: Dynamic Headspace (DHS) is an extraction technique used to concentrate analytes from a solid or liquid sample. The sample is heated and agitated in a sealed vial and the headspace above the sample is purged onto a solid sorbent tube. The solid sorbent tube is then thermally desorbed into the gas chromatograph.

b) Solid Phase Micro-Extraction (SPME)

As solid-phase microextraction is a solvent-free fully automizable sample preparation technique. This is the most sparing for sensitive components and the most time-saving method for the rapid determination of the aroma compounds composition in marjoram, caraway, sage, and thyme.

c) Supercritical Fluid Extraction (SFE)

It is an innovative, clean and environmental friendly technology with particular interest for the extraction of essential oil from plants and herbs. Supercritical CO_2 is selective, there is no associated waste treatment of a toxic solvent and extraction times are moderate.

d) Phytosol Extraction

Phytosol extraction is a new portable cold extraction technique without heat. The aim of the work is the extraction of the volatile components by Microwave-assisted solid phase micro-extraction and Phytosol extraction.

e) Simultaneous Distillation Extraction (SDE)

Simultaneous distillation-extraction using the Likens-Nickerson apparatus is a convenient technique used to isolate volatile organic compounds from complex liquid matrices. The technique combines steam distillation with solvent extraction.

f) Microwave Distillation

Solvent free microwave extraction is an original combination of microwaves and dry distillation. The apparatus is relatively simple. The isolation and concentration of essential oils are performed in a single stage. This process thus frees essential oil which is evaporated by *in situ* water of the plant material.

g) Membrane Extraction

The extraction process involves a partitioning of the analytes from the sample into the organic liquid impregnated in a membrane, followed by diffusion through the membrane into the bulk of the acceptor solution.

Uses of Essential Oils

- Essential oils are widely used in perfumes, personal hygiene products and in aromatherapy including the inhalation, massage, masking agent to avoid the unpleasant odor in the textile industries, paint and plastic industries and pharmaceuticals formulations.
- Essential oils have good medicinal applications and used in the treatment of different diseases including the infectious diseases, depression, anxiety, act as the antifungal, antimicrobial, anticancer and wound healing; they are also used in cosmetics and perfume industries
- Essential oils are primarily used in natural cosmetics due to their anti-inflammatory, antibacterial and antiviral properties.
- Aromatic essential oils (EO) are important plant extracts that have attracted great interest for their various biological activities and their potential to replace chemical preservatives in the field of food preservation. Moreover, the volatility of essential oils has further expanded their applications in antibacterial packaging (AP) systems.

Suggested Readings

Adlard ERK Hüsnü Can Başer and Gerhard Buchbauer (Eds.): Handbook of Essential Oils. Science, Technology and Applications. Chromatographia. 2016; 79(11-12):791.

Davy Bradley 2022 *Essential Oils Encyclopedia,* Chennai: Notion Press

Gulzar Ahmad Nayik and Mohammad Javed Ansari 2023 *Essential Oils: Extraction, Characterization and Applications,* Cambridge: Academic Press

Husnu K Can Baser and Gerhard Buchbauer 2010 *Handbook of Essential Oils: Science, Technology and Applications,* New York: CRC Press

Mozaniel Santana De Oliveria 2020 *Essential Oils – Bioactive Compounds, New Perspectives and Applications,* London: Intech open

Valerie Ann Worwood 2016 *The Complete Book of Essential Oils & Aromatherapy,* Novato, California: New World Library

CHAPTER 7
Tree Borne Oil Seeds

M.P. Divya and S. Krishnamoorthi
Tamil Nadu Agricultural University, Coimbatore

Tree-borne oilseeds are seeds that are borne by trees and are rich in non-edible oils. They are mainly generated by perennial species. As India grapples with its rising imports of vegetable oil, imports account for over 67% of current demand. Non-edible oil seeds are the main source of bio-diesel in India and the technology to make it locally available. Bio-diesel is non-toxic, biodegradable and sulphur-free. The oxygen level is around 10%, which results in superior CO, HC, and PM emission characteristics. Biodiesel has a manufacturing capacity of about 20 million ton per year. The estimated potential ranged from 0.1 to 20 million tonnes with 20 to 25% of that being used. In India, there are over 150 non-edible tree borne oilseeds (TBOs) including sal (*Shorea robusta*), neem (*Azadirachta indica*), mahua (*Madhuca longifolia*), karanj (*Pongamia pinnata*) and kokum (*Garcinia indica*), jatropha (*Jatropha curcas*), kusum (*Schleichera oleosa*), pilu (*Salvadora oleoides*), prinsepia (*Prinsepia utilis*), alexandrian laurel ball tree (*Calophyllum inophyllum*), bitter apple (*Citrullus colocynthis*) and jojoba (*Simmondsia chinensis*). The oil content of these species ranged from 21 to 73 percent (Table 7.1). The majority of these TBOs are spread throughout forest and non-forest areas, with only 20% of the potential being crushed and utilized and that of low quality, due to a lack of awareness of their usage, collection and marketing among other factors (Fig. 7.1).

On the basis of several characteristics such as hardiness, ease of handling and excellent development in a variety of agro climatic situations, *Jatropha curcas* and *Pongamia pinnata* have been found to be the best ideal for biodiesel production among the many Tree Borne Oilseeds (TBOs). These species may be grown in a variety of soil and climatic conditions and the oil from these species is ideal for biodiesel production.

On December 24, 2009, the Union Cabinet approved the National Policy on Biofuels and its implementation. It also approved the formation of a National Biofuel Coordination Committee, which provide policy advice and coordination and will be chaired by the Prime Minister. The Ministry of New and Renewable

Energy has been recognized as the coordinating Ministry for biofuel development and use, with particular tasks assigned to other Ministries.

Fig. 7.1 Seeds of various TBOs

Salient Features of the National Biofuel Policy

- Non-edible oil seeds from waste lands, degraded lands and marginal areas will be used to produce bio-diesel.
- A target of 20% biofuel blending by 2017 has been proposed, for both biodiesel and bioethanol.
- To offer equitable prices to growers, a minimum support price (MSP) for non-edible oil seeds would be announced and periodically revised.
- The Minimum Purchase Price (MPP) for bioethanol and biodiesel would be released and revised on a regular basis.
- Research, development and demonstration will receive a major boost, with an emphasis on plantations, biofuel processing and production including second generation biofuel and financial incentives, such as subsidies and grants.
- A National Biofuel Fund might be considered if it becomes required.

Table 7.1 Seed yield and oil content of important TBOS

Sl. No	Scientific Name	Seed yield/ tree/ year (Kg)	Oil content (%)
1.	*Aleurites fordii*	80	57 to 80
2.	*Azadirachta indica*	35 -50	40 to 45
3.	*Calophyllum inophyllum*	12 - 100	70 to 75
4.	*Garcinia indica*	35	23 to 26
5.	*Jatropha curcas*	3.0 - 6.5	45 to 50
6.	*Madhuca longifolia*	60 - 80	35 to 50
7.	*Pongamia pinnata*	9 - 90	35 to 40
8.	*Prinsepia utilis*	30 -55	65 to 71
9.	*Prunus armeniaca*	35 - 40	40 to 50
10.	*Salvadora oleoides*	10 -15	30 to 35
11.	*Schleichera oleosa*	21 - 28	30 to 34
12.	*Simaroua glauca*	20 - 25	60 to 70
13.	*Simmondsia chinensis*	2 - 3	38 to 68

Processing of TBOs

i) Handling and storage of TBOs

To reduce deterioration and preserve good quality of contained oil and meal, proper handling and storage of oil-containing products is important. Under the right condition, whole, intact, low-moisture oilseeds (approximately 8-10 percent moisture) can be preserved for a long period. Respiration or carbon dioxide evolution, as well as heat generation by oxidation reactions, accompany the deterioration of oilseeds. Seeds that are sound and unbroken may release lesser than 10 cm^3 of carbon dioxide per grams of seed per day, whereas damaged, high moisture seeds may release 50cm^3 or more per grams of seed per day. Furthermore, heat produced by oxidation reactions can raise the temperature of stored seed, hastening deterioration and potentially charring the seeds.

Native enzyme activity in oilseeds, insect and mite infestation and microbial activity during storage are all major factors that affect oil and meal quality. Seeds with high moisture content (more than 14-15 percent moisture) have a negative impact on oil and seed quality. Microbial proliferation and enzyme activity in high-moisture seeds can speed up oil splitting, or acid formation. Free fatty acids make up roughly 0.5 percent of the oil in ripe seeds. The acidity of oil can be substantially higher if seeds are damaged manually or by frost, or if they become moist during harvest, processing and storage.

Another crucial issue during the storage of high-moisture seeds is sprouting, which is considered a detrimental factor. When compared to sound seeds, sprouted seeds may have less oil and more free fatty acids. To keep the seeds cool, most oilseed storage bins have aeration ducts and air fans. Because oxidation reactions are aerobic processes, a low oxygen environment in storage bins aids in reducing oxidation and quality degradation. Because mature seeds have reduced activity of oil-splitting enzymes, they can be stored for longer than immature seeds. Storage of harvested oilseeds properly may also contribute to desired modifications.

ii) Cleaning

Before storing oilseeds, plant stems, twigs, leaves and other material must be removed. Such substance could disintegrate and induce heating in stored seed mass, lowering the quality of the oil and meal. A combination of rotating or vibrating coarse screens, reels and aspiration is commonly used to separate foreign elements from seeds.

iii) Seed drying

Oilseed moisture content is often controlled in order to reduce storage degradation and increase downstream processing efficiency.

iv) Dehulling

Prior to oil extraction, most oilseeds must be removed from their outer husk or shell. Shelling or decorticating is other terms for dehulling. Dehulling improves oil production efficiency, extraction equipment capacity and expeller wear by removing the abrasive husks. Hulls reduce total oil yield by absorbing or holding oil in the pressed cake if they are not removed. Wax and colour chemicals found in the hulls also wind up in the extracted oil.

v) Size Reduction and Flaking

Before oil extraction, oilseeds are reduced in size to make hull removal, heating, drying and flaking easier. Prior to solvent extraction, oilseeds can also be flaked. Flaking causes the cellular structure of the seed to tear, reducing the duration that the solvent must travel to reach the oil within the cells. The seeds are stretched and flattened and pass through the flaking mill. Flake thicknesses typically ranged from 0.01 to 0.015 inch. (0.25-0.37 mm). Oilseed flaking also increases surface area, which allows for more contact between the solvent and the seed during the solvent extraction process. To avoid oil quality deterioration, oil from cracked or flaked seeds should be extracted as soon as possible.

vi) Tempering

Oilseeds are cooked or tempered to denature proteins, release oil from the cells and inactivate enzymes. In less than 5 min, it is preheated to 68-122° F and contacted with 248° F active steam. To remove oil from cooked seeds, they are pressed immediately. Cooking also increases the flexibility of the seeds, allowing for easier pressing.

Extraction Methods

a) Mechanical Pressing

TBOs seeds are placed in a container and squeezed with a press or screw until the oil is extracted. The pressure is gradually increased from 138 to 412 bar. When the oil stopped flowing out, the mechanical extraction procedure is completed in 25 min. and the oil is collected in a drum. This filtration is done to remove any undesired particles that may have remained in the extracted oil, resulting in pure oil (Fig. 7.2).

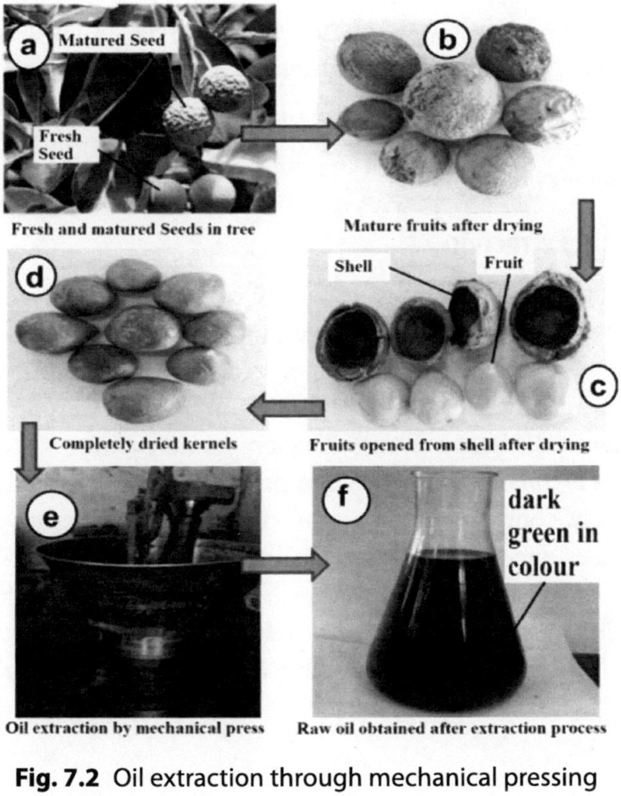

Fig. 7.2 Oil extraction through mechanical pressing

b) Steam Distillation Extraction

The oil is extracted using steam and high pressure in this procedure. The oil flow is increased by heating the kernels with steam, then squeezing them under high pressure. The kernels contain the most of the oil. The extraction process is made easier using this approach. The dried seeds are placed in the steam boiler during the steam distillation process. The steaming causes them to swell, making oil squeezing much easier. This steaming process is accompanied by an increase in boiler pressure which extracts the oil from the seeds. Steam distillation is a separation technique that uses steam to separate a mixture of heat sensitive components. The approach is based on decreasing the partial pressure of volatile components by changing the boiling point. Steam distillation is used in petroleum refineries to extract volatile plant chemicals that are used in the fragrance and flavouring industries. There are two parts to the steam distilled content viz., essential oil and distillation water. The secondary essential oil is recovered from the steam distillate using solvent extraction.

c) Solvent Extraction

The solvent, often hexane, is combined with the seeds after they have been oven dried and pressed in the solvent extraction process. It operates on the counter current and moving bed concepts, allowing for good solvent penetration and percolation for absolute oil extraction. The extracted oil is collected by evaporating the solvent and recovering it afterwards.

d) Super Critical Extraction

The most effective and efficient technique to extract valuable constituent botanicals is through the use of supercritical fluid extraction. This is the process of separating one component or extractant from another or the matrix utilizing supercritical fluids as the extracting solvent, which is carbon dioxide. Carbon dioxide solvent is the king of plant extraction solvents. The temperature and pressure for supercritical CO_2 extraction are above the critical temperature of 31°C and 74 bar. Supercritical fluids are highly compressed gases with fascinating properties that combine the features of gases and liquids. In normal solvents, supercritical fluids can lead to reactions that are difficult to achieve. It is a quick operation that takes 10 to 60 min to complete. By gradually releasing pressure, a supercritical extraction fluid can be separated from analytes, leaving almost minimal trace and yielding a purified residue (Fig. 7.3).

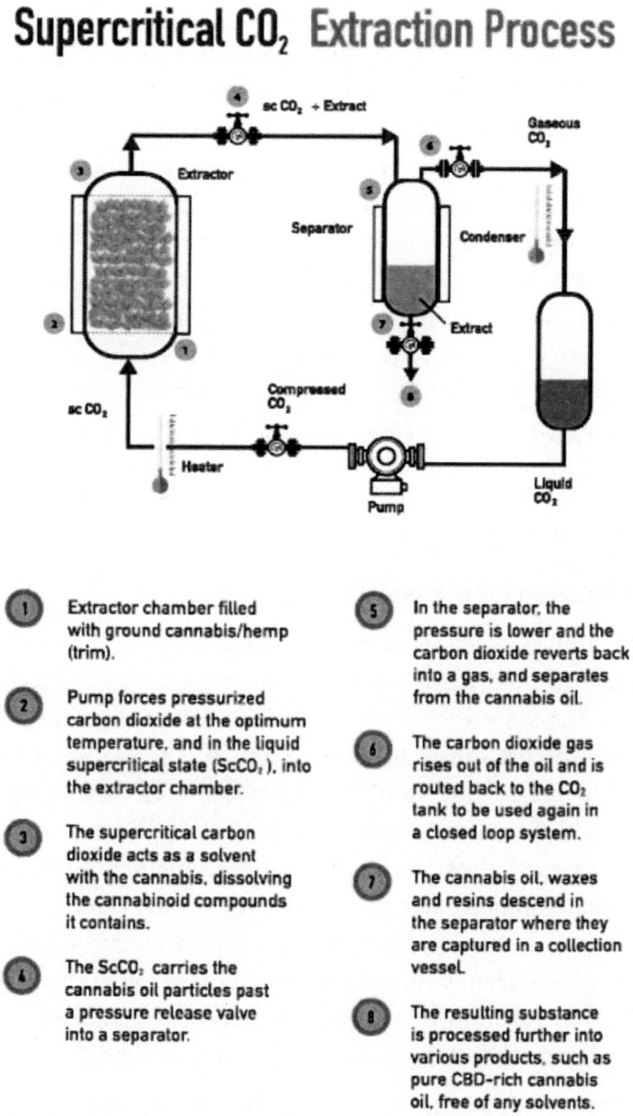

Fig. 7.3 Super critical extraction meethod

e) Aqueous Extraction

The most common approach is aqueous extraction, which is the simplest and most widely used. It entails crushing or grinding the seeds in water, straining the extract into a thin fabric and collecting the liquid. This oil extract can be used as a pesticide spray without any modifications.

Biofuels

Biofuels are the potential renewable and alternative to petroleum-based fossil fuels. With over 100 different tree species producing seed oil appropriate for biodiesel. India has a lot of potential. These oil bearing tree species can be planted in non-agricultural areas, so they don't compete for food and fodder with livestock. With the launch of both the National Biofuel Policy (2009) and the National Agroforestry Policy (2014) in the country, the first policy aims to replace fossil fuels with biofuel to the tune of 5% by 2012, 10% by 2017 and above 10% beyond 2017, while the second policy aims at integrated land use options for livelihood, environment and energy security. Some of the most significant TBOs, such as jatropha, karanj, mahua, simarouba, etc. are successfully integrated into agroforestry systems. Biodiesel is produced by transesterification by alkali process as it results in high purity and yield of the biodiesel product in a short time.

Biofuel Extraction Method

i) Transesterification for Biodiesel Conversion

Transesterification is a chemical reaction used for the conversion of triglycerides (fats) contained in oils (feedstocks) into usable biodiesel. Biodiesel produced by the process of transesterification has a much lower viscosity, making it capable of replacing petroleum diesel in diesel engines. The biodiesel production method applied worldwide at industrial level is the alkalosis (transesterification) of triglycerides which are the main component of biofuel. The most common type of esters is methyl esters mainly because methanol is usually the cheapest alcohol. The reaction is catalysed by bases, acids and enzymes and can take place either at low or high temperatures. The produced methyl esters of fatty acids are characterized as biodiesel while glycerol is received as a reaction by-product. The transesterification reaction of triglycerides where R, R' and R" are long chains containing carbon and hydrogen atoms (Chain fatty acids).

In this process, the alcohol, the catalyst and the oil are mixed in a reactor and stirred for 1 h at 60°C approx. Smaller capacity plants often use non-continuous reactors (Batch reactors), but most large plants (those generating more than 4 million litre/year) reactors are using continuous and complete mixing. The reaction generally includes two steps. During the first one, it is added 80% alcohol and the catalyst in the oil. Then, the reactor output stream is led to the removal of glycerin before entering the remaining 20% of alcohol. After the reaction, the glycerol is separated from the methyl esters. Due to the low solubility of the glycerol esters, their separation can be accomplished easily and quickly using precipitating tank or simple centrifugation. The amount of methanol that has not reacted, acts as a solvent thereby delaying the separation of glycerol and methyl. However, excess methanol is usually not removed from the reaction stream until a complete separation of glycerol and methyl esters has been achieved, because the direct reaction of transesterification of products (higher conversion). Often, after the completion of the transesterification, water is added to the reaction mixture to facilitate the glycerol separation.

It is possible to make oil react with methanol in the absence of a catalyst so as to minimize the need of washing with water. In this case, however, high temperatures are required and larger excess amount of methanol. Also, there has been difficulty in replication kinetics of the reaction, which has been attributed to various catalysts on the surface of the reactor. Moreover, it has been found that at high temperature and pressure (about 240°C and 90bar respectively) the transesterification of fats can take place without prior removal or conversion of free fatty acids. For the synthesis of biodiesel, the following materials oil sample, methanol and potassium hydroxide (KOH) are used as a catalyst. Methanol and potassium hydroxide are pre-mixed to prepare potassium methoxide and then added to oil in the reactor with a mixing speed of 400 rpm for 2 h at 50°C. The molar ratio of oil to methanol is 1:10. Finally, the mixture is left overnight to settle forming two layers, viz., biodiesel phase (upper layer) and the glycerin-rich phase (Fig. 7.4)

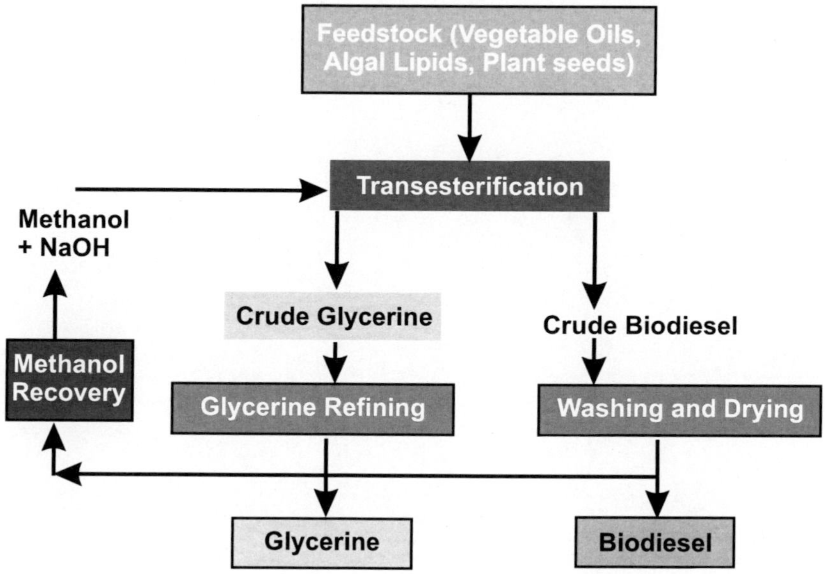

Fig. 7.4 Transesterification process of biodiesel production

Suggested Readings

Chakravarty Sumit Shukla Gopal Dey Amarendra Nath 2012 *Tree-Borne Oilseeds Species,* Saarbrucken, Germany: LAP Lambert Academic Publishing

David M Mousdale 2012 *Introduction to Biofuels,* New York: CRC Press

Gayatri Vaidya 2021 *Textbook of Biofuel Technology,* Saarbrucken, Germany: LAP Lambert Academic Publishing

Markus Schmidt Manuel Porcar Vincent Schachter Antoine Danchin and Ismail Mahmutoglu 2010 *Biofuels,* Weinheim, Germany: Wiley-VCH Verlag GmbH & Co. KGaA

Sundaram Suresh Ramanan 2014 *An Intro Tree Borne Oil Seeds,* Riga Latvia: KS Omni Scriptum Publishing

CHAPTER 8
Lac and Shellac

V. Priyanka and B. Vinothini
Forest College and Research Institute, Mettupalayam

In an agricultural country like India, along with the cultivation of crops, insects are also cultivated. With the need for a little technical knowledge and less time in lac insect rearing, it can be cultivated easily. Lac is nature's gift to human kind as it is natural, biodegradable, non-toxic, odourless, tasteless, non-injurious to health hard resin. It is a resinous exudation from the body of female scale insect. Since, lac insects are cultured on host trees which are growing primarily in wasteland areas, promotion of lac and its culture can help in ecosystem development as well as reasonably high economic returns. It is a source of livelihood of tribal and poor inhabiting forest and sub forest areas.

Lac insect of South East Asia is referred to as *Kerria chinensis*. Lac insect belongs to super family Coccoidea which includes all scale insects. The mouth part of these insects is piercing and sucking type. They can be very destructive to tree making it stunting or killing twigs and branches by draining the sap. There are six genera of lac insects, out of which only five secrete lac and only one, *i.e. Laccifera* secretes recoverable or commercial lac. The commonest and most widely occurring species of lac insect in India is *Laccifera lacca* (Kerr) which produces the bulk of commercial lac.

Biology of Insect in Lac Cultivation

Lac is a type of natural resin that is formed as a result of secretion by the female Indian lac insect, *Laccifera lacca* (Kerr). It belongs to Kerridae family, consists of nine genera, while the number of species reported vary from 87 to 100 species. Two genera are found in India, while genus *Kerria* is the most important and widely exploited insect for lac cultivation in India. Lac insect is a soft-bodied, round tiny creature, which completes its life cycle in four stages *viz*., egg, larva, pupa and adult on host plants within six months. The adult male lac insect lives for a very short duration, such as 3-4 days, while the female lac insects live longer. During the life cycle, this insect suck the sap juices of tree branches through its mouth and the female lac insect secretes lac around the branches of host plants by which stick lac is obtained, thus it plays a major role in the production of lac.

Distribution of Lac Insects

Since the lac insects bloom and feed on certain species of the tropical trees, it is found distributed in South East Asian countries. Lac is currently produced in India, Myanmar, Thailand, Malaya, Lao and Yuan province of China and to some extent in Vietnam and Mexico. It is generally categorized as a 'non-timber forest product'. India and Thailand are main areas in the world, while India has leading position in relation to lac production. Lac cultivation is introduced into Thailand from India. Over 90% of Indian lac produced comes from the states of Bihar, Jharkhand, West Bengal, Madhya Pradesh, Chhattisgarh, Eastern Maharashtra and northern Orissa. Some pockets of lac cultivation also exist in Andhra Pradesh, Mysore, Gujarat and Mirzapur and Sonbhadra district of Uttar Pradesh. In India, lac is cultivated in 50 districts spread over 12 states. In Jharkhand, lac cultivation constitutes the second most important source of rural family income. Due to its diversified uses, demand of lac is increasing globally and price of the commodity is also in increasing trend.

Status of Raw Lac Production

India is the leading lac producer in the world in terms of the production of raw lac, with an annual production of over 20,000 tons. About 80 per cent of the world's total production is in India and 75 per cent of it is exported to over hundred countries, mainly in processed and semi-processed forms. After India, lac is produced more in Thailand. Lac is also produced in Indonesia, parts of China, Myanmar, Philippines, Vietnam and Cambodia, etc. In India, lac production takes place mainly in Chhota Nagpur region of Jharkhand, Chhattisgarh, Madhya Pradesh, West Bengal, Orissa, Uttar Pradesh and Maharashtra. Among the lac growing states, Jharkhand state ranks 1st followed by Chhattisgarh, Madhya Pradesh, Maharashtra and Odisha and the contribution of these five states in national lac production is about 53%, 17%, 12%, 8% and 3%, respectively. This major lac producing five states contribute around 93% of the national lac production.

Host Plants

Lac insects thrive on twigs of certain plant species, suck the plant sap and grow all the while secreting lac resin from their bodies. These plants are called host plants. Although lac insect is a natural pest on host plant, these insects enjoy the privileged position not being treated as pest because of the following reasons viz., i) they yield a useful product, ii)the host plants are economically not so important and iii) the insects cause only temporary and recoverable damage to the host plants.

About 113 varieties of host plants are reported as lac host plant. Out of which, the followings are very common in India.

- *Butea monosperma* (Palas)
- *Zizyphus* spp (Ber)
- *Schleichera oleosa* (Kusum)
- *Acacia catechu* (Khair)
- *Acacia arabica* (Babul)
- *Acacia auriculiformis* (Akashmani)
- *Zizyphus xylopyrus* (Khatber)
- *Shorea talura* (Sal)
- *Cajanus cajan* (Pigeon-pea)
- *Grewia tiliaefolia* (Dhaman)
- *Albizzia lebbek* (Siris)
- *Flemingia macrophylla* (Bholia)
- *Ficus benghalensis* (Bargad)
- *Ficus religiosa* (Peepal)
- *Albizzia lebbeck*(Siris)

Of these host plants, palas, kusum, ber and khair are of major importance, while others are of regional importance. India contributes 95% of commercial production of lac. It is also important to mention that the quality of lac is directly related to the host plant and to the strain of lac insects.

Strains of Lac Insect

In India, lac insect is known to have two distinct strains: i) Kusumi and ii) Rangeeni.

i) Kusumi strain

The Kusumi strain is grown on Kusum or on other host plants using Kusumi brood. The life cycle of lac insects takes about six months; hence, two crops a year can be obtained ie. i) Jethwi (June/July) and ii) Aghani (Jan. /Feb). The Kusumi crop lac is considered superior resin, because of the lighter colour of resin and it fetches better price.

ii) Rangeeni strain

In case of Rangeeni also, two crops are: i) Karrtiki (Oct. /Nov.) and ii) Baisakhi (May/June) can be obtained. The lac of Rangeeni crops is harvested while it is still immature. Aghani and Baisakhi of Rangeeni strain are the main crops contributing about 90% of lac production and the remaining 10% is contributed by Kusumi crops.

Lac Cultivation

Lac cultivation is done by the farmer acquiring a 'broodlac', a stick covered with eggs of the insect that are on the verge of hatching. This broodlac is tied on to the branch of a tree where the insects are intended to infest on. Lakhs of the insects grow into large colonies on the branches of the tree and secrete the resinous product. The resin coated branches of the tree, referred to as sticklac, which are cut and processed to obtain lac. These sticklac branches are crushed and sieved to remove most of the impurities. The sieved powder is then repeatedly washed with cold water in order to obtain a purified product known as seedlac due to it resembling pellet in shape. The remaining impurities in seedlac is removed either by dissolving in solvent or by heating. Major lac cultivation operations consist of six stages i) selection of suitable host plants, ii) inoculation of brood lac, iii) removal of brood lac sticks, iv) natural enemies of lac insect, v) harvesting of lac sticks and vi) scraping of raw lac from twigs.

i) Selection of suitable host trees for lac cultivation

When starting cultivation in new areas having a lac host, the trees should be pruned before infection to ensure good lac production. Selected lac hosts should have the following salient features viz., i) fairly fast growing, ii) lower sap density and iii) well adapted to pollarding. To get soft and juicy twigs in the nutritious trees, pruning of trees is necessary at a certain time so that lac insects can be easily reared.

ii) Inoculation of brood lac

Brood lac is a mature lac from where the young insects are ready to come out within the time specified. To achieve the best result from lac cultivation, the work should be planned in systematic manner. In order to transmit the lacquer insect to the nutritious trees, a bundle of 6 to 9 inches long, 3 to 4 stalks of the lacquer is made, which is placed at several places of the lacquer tree. This operation includes allowing young lac larvae (crawlers) to come out of their mother cells and settle on the host plant.

iii) Removal of broodlac sticks

Generally, the emergence of lac larvae from the brood lac ceases after three weeks. This operation should be done to prevent access of the insect predators and parasitoids of lac insect to new lac crop and to avoid wastage of lac after drying up of stick lac and prevent its falling on the ground.

iv) Natural enemies of lac insect

Lac insect is mainly attacked by two types of natural enemies such as i) parasites and ii) predators. In the case of lac insect, small tiny winged parasites such as *Tetrastichus purpureus* is the most abundant lac associated parasite. They lay their eggs in the lac cells and the larvae (grubs) hatching out feed on the lac insect within its cells. On the other hand, predators which directly involved in the consumption of their host. It is more serious and can damage up to 30-35 % cells in a crop. *Eublemma amabilis* and *Pseudohypatopa pulvera* are the most destructive key pests of lac insect.

v) Harvesting lac crop

Harvesting is the process of collection of ready lac from host trees. It is done by cutting the lac encrusted twigs when the crop is mature. There are two types of harvesting.

Immature harvesting: Lac is collected before swarming and lac thus obtained is known as Ari Lac. The immature harvesting has drawbacks, as the lac insects may be damaged at the time of harvesting. However, in case of Palas lac (Rangeeni lac), it is found that Ari Lac gives better production. Hence, Ari Lac harvesting is recommended in case of Palas only.

Mature harvesting: Lac is collected after swarming. The lac obtained is known as mature lac. A yellow spot develops on the posterior side of lac cell towards crop maturity. This spot spread forwards until it covers half of the cell. Cutting of twigs for harvest can be done at any time between the stages while yellow spot occupies one third to one half of the cell area. The harvesting periods of different crops are different. The Karrtiki crop is harvested in October/ November; Baisakhi in May/June; Aghani in January/February and Jethwi in June/July. When the nymphs have escaped from the brood lac, left is the stick lac or phunki lac. These sticks should be tied in bundles and immersed in water, preferably running water for 3- 4 days, keeping them well under water with help of heavy stones. The stick lac should then be kept in shade for drying. The raw lac should be scraped while sticks are still moist. Estimated yield obtained per tree in India ranged from 6 to 10 kg for kusum, 1.5 to 6 kg for ber and 1–4 kg for dhak. The insect life cycles can produce two stick lac yields per year, though it may be better to rest for six months to let the host tree recover.

vi) Scrapping of raw lac from twigs

Scraping is a process in which encrusted lac resin is removed from lac host stick. After harvesting of matured lac and sometimes immature lac is needed to be scraped as primary processing for long time storage. This scrapping is done with the help of a scraping knife / crusher for different applications in the processing area is given below.

Local practice: Lac cultivation is simple and does not need any large investment and requires only part-time attention. The local practice in lac cultivation has some disadvantages like i) the same host plants are continuously exploited without giving rest for recoupment, ii) only natural inoculation occurs and iii) partial harvest is done leaving few branches untouched for auto inoculation of next crop and no pruning is done. As a result of the defective local practices, host trees loose the vigour and unable to throw out new succulent shoots and in course of time, the trees become weak and die. The self-inoculation leads to heterogeneous infestation of nymphs, which results in wholesome mortality of brood in seasons of extreme heat.

Improved practice: Sustained production of lac and steady returns can be achieved by adopting improved method of cultivation. The underlying principle in improved method of lac cultivation is to provide much needed rest to the host plants after a harvest has been taken. For this purpose, coupe system of lac cultivation is adopted. As the term coupe means a chamber, the host plant trees are divided into coupes *i.e.*, groups that consist of certain number of trees. In practice, only few numbers of trees in a coupe are inoculated. Following harvest, these trees are made to rest and recoup the lost vigour, while other trees are ready with succulent twigs for inoculation. Thus, in a coupe system, alternate groups of trees are put to lac cultivation. Full inoculation and full cropping are the rule under this system. The following considerations are desirable in improved lac cultivation which include superior breeds of lac insect, providing proper rest to host plants, use of good quality brood lac in appropriate quantity, post-harvest management of lac crop, host plant management and lac pest management. Timely harvesting of mature crop and proper inoculation will reduce the risk of loss of lac insect to a large extent.

Propagation of Lac Insects

Propagation means the spread of lac insects on the same or different host plants. This is done by inoculation of newly hatched (brood) nymphs. Inoculation is of two types.

a) Natural or Self Inoculation

This type of inoculation occurs naturally. It is very simple and common process, when the swarmed nymphs infect the same host plant again. Natural inoculation, being repeated on the same host, makes host plant weak and thereby, nymphs do not get proper nutrition. Also in natural inoculation, it is not sure that uniform sequence of inoculation takes place. Therefore, natural inoculation should be discouraged.

b) Artificial inoculation

The main idea behind the artificial inoculation is to check the drawbacks of natural inoculation. In this method, the host plants are first of all pruned in January or June. It induces host plants to throw out new succulent twigs. Only light pruning should be carried out. In artificial inoculation, brood twigs are cut with a size of 20 - 30 cm in length. Then, the cut pieces of brood twigs are tied to fresh tree twigs in such a way that each stick touches the tender branches of trees at several places. The nymphs swarm from brood and migrate to tender and succulent twigs and infest them. Following swarming, the brood twigs should be removed from the host plant, as this would decrease the chance of pest infestation. The precautions to be taken in artificial inoculation are listed below.

- Fully mature and healthy brood free from pest infestations should be taken.
- Brood meant for inoculation should not be kept for long and used immediately after crop cutting.
- Tying of the brood lac stick should be done securely on the upper surface of branches. This will prevent falling of twigs and provide full contact for quick and easy crawling of the nymphs. One should keep a watch on the brood lac dropping down.
- Sometimes due to bad weather, swarming of nymphs from brood is prevented. Hence, the room storing brood lac sticks is moderately heated to 20°C to induce swarming and then sticks are tied.
- Generally, cultivation of Kusumi in Rangeeni area and vice versa should be avoided. Brood lac from a particular host used year after year is likely to deteriorate in quality. Therefore, alternation of brood and host give production of a better quality of brood lac.

Composition of Lac

The major constituent of lac is the resin. Lac resin is a polyester complex of straight chain hydroxy fatty acids of $C_{14} – C_{18}$ carbon chain (such as Aleuritic acid, Butolic acids), mono- and di – hydroxy acids along with hydroxyterpenic acids. Other constituents are dye, wax, sugar, proteins, soluble salts, sand, woody matter, insect body debris, etc. Lac wax is a mixture of anthroquinoid derivatives. Composition of lac includes resin: 68 to 90%, dye: 2 to 10%, wax: 5 to 6%, mineral matter: 3 to 7%, albuminous matter: 5 to 10% and water: 2 to 3%

Lac Processing

a) Stick lac

Following harvest, lac encrustations are removed from the twigs of host plant by scraping. The raw lac thus obtained is known as raw or crude lac or scraped lac or stick lac. This crude lac consists of resin, encrusted insect body, lac dye, sand and twig debris. The freshly scraped lac contains a lot of moisture and usually left to dry. The quality and value of stick lac depend very much upon variety of factors, *viz.* host tree, climate, whether the crop is harvested before or after emergence of larvae and the method of drying and storage. The stick lac cannot be stored for longer duration, as the lac has tendency to form lump and there is loss in quality of lac. High moisture content is responsible for lump formation. The optimum moisture content has been identified to be 4% for storage of stick lac to avoid lump formation. It is recommended to store the stick lac on floor in layers less than 30 cm in height and racked frequently. If stick lac is converted into seed lac, it can be stored for longer duration like food grains (Fig. 8.1).

b) Seed lac

The stick lac is crushed and sieved to remove sand and dust. It is then washed in large vats again and again to break open the encrusted insect bodies, to wash out the lac dye and twig debris. Decaying bug bodies turn the water a deep red that is processed further to get the byproduct lac dye. The remaining resin is dried, winnowed and sieved to get the semi refined commercial variety product called seed lac. The dusty lac eliminated by sieving is refuse lac known as molamma. The seed lac is in form of grain of 10 mesh or smaller and yellow or reddish brown in colour in general appearance. Adhering impurities on the grains of seed lac may be 3 - 8%.

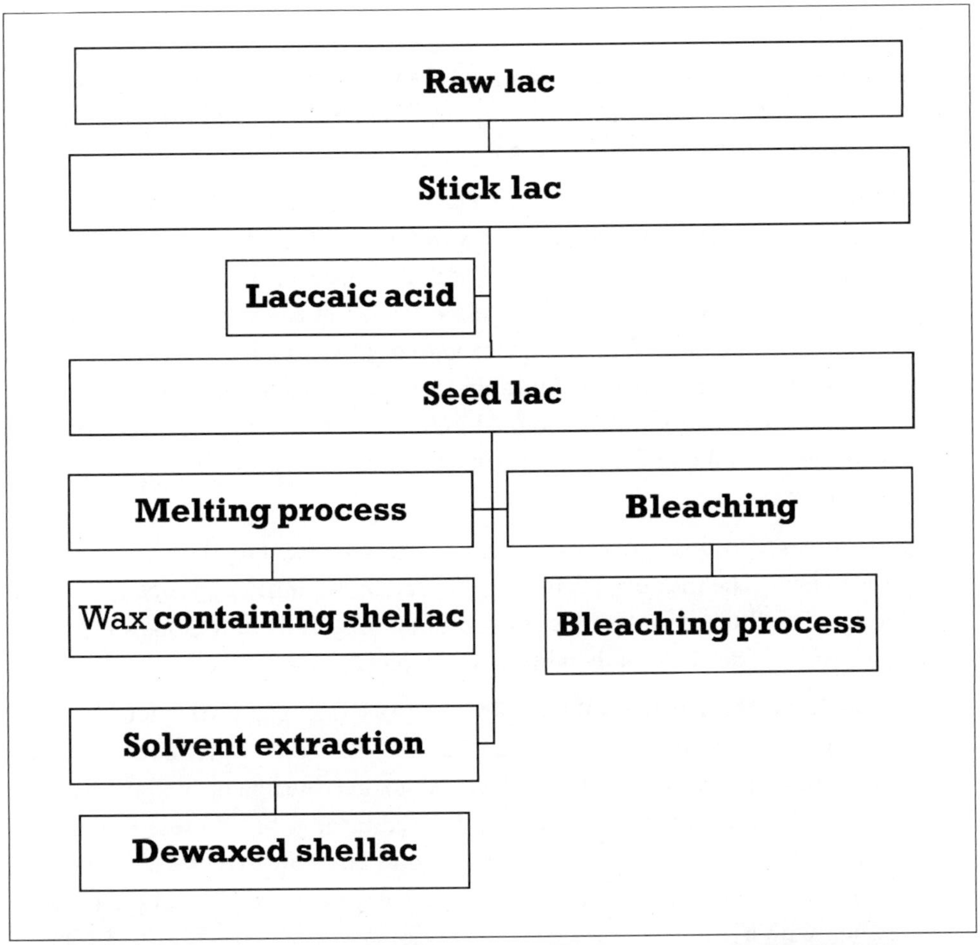

Fig. 8.1 Lac Processing

c) Shellac

The shellac is the name of finished product and is commonly used across the world. Seed lac is processed into shellac by any of the three methods viz., handmade country process, heat process and solvent process.

Handmade process: Traditionally seed lac is processed by hand. The seed lac is filled into long sausage shaped cloth bag of about 2 inch. diameter and 30 feet long. The long bag is passed gradually in front of a charcoal fired hearth hot enough to melt the lac. By twisting the bag, molten lac is squeezed out through cloth. The residue left inside cloth bag is another variety of refuse lac known as kirilac. The molten filtered mass is stretched into sheets approximately 0.5 cm thick and thinner by skilled work man with the help of glazed ceramic cylinder. Alternatively, the molten mass is allowed to solidify in the form of discs and then it is called as 'button lac'.

Heat process: In this process of manufacturing of shellac, the seed lac is melted by steam heat. The molten soft lac is squeezed through filter by means of hydraulic pressure. The filtered molten lac is drawn into long and continuous sheets with the help of roller. The sheet is then broken into pieces called flakes.

Solvent processes: If the solvent process is used to purify the semi refined lac, dewaxed and decolorized shellac can be obtained as end product. The normally amber colour resin can also be bleached to get bleached shellac. Seed lac is dissolved in a refrigerated alcohol and filtered through filter press to remove wax and impurities. The colour may be removed to any required standard by charging with the activated carbon and then alcohol is recovered. The molten shellac is stretched with a roller.

Lac Products and its Uses

a) Lac dye

Lac dye is a mixture of anthroquinoid derivatives. It is traditionally used to colour wool and silk. Its colour varies between purple red, brown and orange often depending upon the mordant used. It is used in food and beverages industry for colouring. In recent past, lac dye has been replaced by synthetic dye. But, nowadays with increasing stress and awareness on use of eco-friendly and safe material particularly associated with human contact and consumption has made revival of great demand of lac dye as a colouring material.

b) Lac wax

Lac wax is a mixture of higher alcohols, acids and their esters. It is used in polishes (shoes), floor, automobiles, etc., food and confectionary and drug tablet finishing, lipsticks and crayons.

c) Shellac

Shellac is a natural gum resin, a nature's gift to the mankind and is used in over 100 industries. It is natural, nontoxic, physiologically harmless and edible resin. Shellac is a hard, tough, amorphous and brittle resin containing small amount of wax and a substance responsible for its characteristic pleasant odour. The lac resin is not a single chemical compound, but an intimate mixture of several components. Shellac is slightly heavier than water. Its natural colour varies from dark red to light yellow. Shellac is insoluble in water, glycerol, hydrocarbon solvents and esters, but dissolves readily in alcohols and organic acids. The solvent most commonly employed to dissolve shellac is methylated spirit. Usually the milder alkalis, ammonia, borax and sodium carbonate can also be employed to prepare aqueous solutions. Normal wax content of shellac is 5% which is insoluble in alcohol. It is soluble in n-hexane, pure turpentine and other hydrocarbon oils. The properties of shellac are given hereunder.

- Shellac is thermoplastic
- It is UV-resistant
- It has excellent dielectric properties, dielectric strength, low dielectric consent, good tracking resistance, etc.
- It has excellent film forming properties. Its film shows excellent adhesion to wide variety of surfaces and possesses high gloss, hardness and strength.
- It is a powerful bonding material with low thermal conductivity and a small coefficient of expansion.

Uses

- Shellac is used in fruit coatings, e.g. for citrus fruits and apples, parting and glazing agents for sweets, marzipan, chocolate, etc. It is also used as binder for foodstuff, stamp inks, eg. for cheese and eggs
- It is used as binder for mascara, nail varnish additive, conditioning shampoo, film forming agent for hair spray, micro-encapsulation for perfumes
- It is used for enteric coatings for tablets and as odour barrier for dragees
- It is used in the manufacturing of photographic material, lithographic ink and for stiffening felt and hat material

- It is utilized in preparation of gramophone records
- Jewelers and goldsmiths use lac as a filling material in the hollows in ornaments
- It is also used in preparation of toys, buttons, pottery and artificial leather
- It is also used commonly as sealing wax. With increasing environmental awareness of consumers, this natural and renewable raw material is being used increasingly in the development of new products
- It is used in seasoning leather care products
- It is used as a binder for flexographic printing inks for non-toxic printing of food packaging
- It is used in primers, polishes, matt finishes as wood treatment.
- It is used as stiffeners in textiles
- It is used in insulation, capping and lamination in electricals
- It is used as binder for grinding wheels
- It is used as binder for inks and water colours and micro encapsulation for dyes.

d) Bleached shellac

Bleached shellac is non-toxic, physiologically harmless (edible) and is widely used in the food industries, food packaging and allied industries. Apart from the above, bleached shellac is also used for its qualities i.e. binding, adhesive, hardening, gloss, odourless, fast drying and extending shelf life

Uses

- It is used as primer for plastic parts and plastic film in paints
- It is used as primer for aluminium and aluminium foils in aluminium industries
- It is used in flexographic printing inks
- It is used for coating of pills, tables and gel capsules in pharmaceuticals
- It is used as coating of confections, chewing gums, marzipan chocolates, nutties, jelly and coffee-beans in confectionery
- It is used as binder for food marking and stamping inks and binder for egg coating
- As barrier coating for processed food, vegetables, fruits and dry flowers
- It is used as textile auxiliaries and felt hat-stiffening agents
- It is used in hair spray, hair and lacquers, hair shampoos and binder for mascara in cosmetics

- It is used in wood finishing as binder for wood coatings and wood stains and as filler/sealer for porous surfaces and cracks
- It is used in antique frames for paintings and wood polish (french polish)
- It is used as binder for fireworks, matches, etc and used in coating of magnesia in fireworks and pyrotechnics
- It is used as binder for lamp cements in electricals
- It is a binder for insulation materials, serves as additive to moulding compounds in electronics, mass coating for print plates and is adhesive for Si-cells.
- It is binder for additive of grinding wheels
- It is a primer for plastic parts and films in plastics
- It is an additive to natural rubber
- It is used in leather auxiliaries

e) Dewaxed bleached shellac

Dewaxed white shellac is used in the same way as any other grade of shellac. The major difference between this shellac and the others is that it is a bit harder, shines a bit brighter and completely free from wax. Bleached lac has super characteristics and qualities i.e. adhesive, binding, hardening, gloss, odourless. It has good film forming properties, a high gloss and excellent adhesion to various substrates including the human hair. It is non-toxic and physiologically harmless.

Uses

- It is used in coating of fruits and vegetables
- It is used in coating in tablets and capsules
- It is used in coating in confectionary
- It is used in coating in aluminium foil and paper
- It is used in coating in cosmetic industry
- It is used in hair sprays (pump sprays or aerosol sprays, hair setting lotions, hair shampoos, mascara, eyeliners, nail polishes, lipsticks, micro encapsulation by coacervation of fragrances and perfume oils in cosmetics
- It is used for coating of confections, chewing gum, candles, cakes, eggs, citrus fruits and apples and printing inks for eggs and cheese in food

e) Aleuritic Acid (Shellac Aleuritic Powder)

Aleuritic acid (9, 10, 16-Trihydroxypalmitic Acid) obtained from shellac by saponification, is a unique acid containing three hydroxyl groups of which two are of adjacent carbon atoms. Technical grade aleuritic acid (purity 99%) is a slight yellow and almost odourless solid.

Uses

- There is a continuous growing demand of aleuritic acid in the field of perfumery and pharmaceuticals due to its excellent starting material for the synthesis of civetone, ambrettolide, isoambrettolide, etc, which have the musk like odour. Civetone is obtained from shellac aleuritic acid. It is used for manufacturing of perfumes and is very much in demand by the perfume manufacturing companies in France, Italy, Germany, USA, etc.
- Synthesis of glucose mono aleuritate (a non-toxic non-hemolytic water-soluble compound) in medicine as an isocaloric substitute for dietary tripalmitin.
- Preparation of plastics with good adhesive properties by the condensation of aleuritic acid with pithalic andydride and glycerin, rosin, etc.
- Aleuritic acid esters used in the preparation of lacquers and plastics.

Suggested Readings

Edward Hicks 1961 *Shellac: Its Origin and Applications,* Gloucester: Chemical Publishing Company

Ghorai N 2020 *Lac-Culture in India,* New Delhi: Daya Publishing House

Krishna Murthy T 2010 *Minor Forest Products of India (Non-Timber Forest Products of India,* Hyderabad: BS Publications/BSP Books

Kumar K K R Ramani and K K Sharma 2002 Recent Advances in Lac Culture, ILRI, Namkum, Ranchi,

Patrick Moore Glover 2009 *Lac Cultivation in India: Being a Second and Revised Edition of "A Practical Annual of Lac Cultivation,* Ranchi: Indian Lac Research Institute

Sarkar PC 2011 Application of lac: past, present and emerging trends: in, advance in lac culture. Indian Institute of Natural Resin and Gum, Namkum, Ranchi: 252-258.

Sharma KK and R Ramani 2010 Recent advances in lac culture. IINRG, Ranchi: 1-391.

CHAPTER 9
Medicinal Plants and Drugs

M.P. Divya and P. Hemalatha
Tamil Nadu Agricultural University, Coimbatore

Medicinal plants can be defined as plants that possess therapeutic properties or exert beneficial pharmacological effects on the human or animal body. People have been looking for remedies in nature to cure their illnesses since ancient times. As the reasons for the use of various medicinal plants for the treatment of specific ailments were revealed over time, the use of medicinal plants shifted from an empiric to an explicatory framework. Plants have been the source of treatment and prophylaxis until the invention of iatrochemistry in the 16th century. Nonetheless, the decreasing efficacy of synthetic medications and the growing number of contraindications to their use has brought natural drug use back into the spotlight. The World Health Organization has estimated that the annual demand for medicinal plants is $14 billion and is expected to rise to more than $5 trillion by 2050.

According to the World Health Organization, 80 percent of people worldwide use herbal medicines for some part of their primary health care. Around 21,000 plant species have the potential to be used as medicinal plants. An approximately three-quarter of the world's population get their health care primarily from plants and plant extracts. More than 30% of all plant species have been used for medical reasons at one time or the other. Plant medications are expected to account for up to 25 % of total drugs in developed countries like the United States, while they account for up to 80 % in developing countries like India and China. As a result, medicinal plants are far more important economically to countries like India than the rest of the globe. These countries offer two third of the plants used in modern medicine and the rural population's health care system is based on indigenous medicine.

Tropical medicinal plants provide the basis for about 10 % of today's drugs, including quinine, curare and a variety of steroids. The therapeutic impact of plant based drugs is dependent on a variety of extractives found in the plants. These could be glucoside alkaloids or their derivatives. They are the breakdown products of the plant's metabolism. They are extracted from plant tissues using a variety of methods and their therapeutic efficacy is based on the physiological effects they have on the human body. They are derived from plants such as trees, shrubs and herbs. They are solids, liquids, infusions or dust derived from fruit, flower, leaf, stem or root. The majority of medicinal plants are hand harvested and dried locally. Before the medications can be commercialized and extracts obtained, they must be treated. Pain relievers, hypnotics and stimulants are all possible drugs. Antiseptics and astringents are drugs that have the ability to constrict or contract biological tissues, bring together cut or torn tissues and stop bleeding.

Classification of Plant Drugs

The plant drugs are classified based on the plant parts into a) root drugs, b) bark drugs, c) flower, fruit and seed drugs and d) leaf drugs (Fig. 9.1).

a) Root drugs

b) Bark Drugs

c) Flower, Fruit and Seed drugs

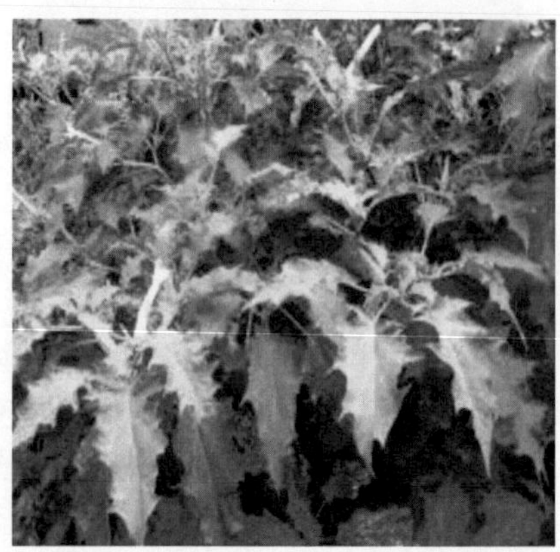

d) Leaf drugs

Fig. 9.1 Classification of Plant Drugs

a) Root Drugs

Root drugs comprise approximately 150 of the 900 species used commercially in Indian medicine. Root drugs are more effective and essential. Tuberous root drugs are ubiquitous.

Very popular root drugs are *Asparagus racemosus* (Shatavari), *Withania somnifera* (Ashwagandha), *Boerhavia diffusa* (punarnava), *Chlorophytum borivilianum* (safed musale), *Tribulus terrestris* (gokhru), *Salacia chinensis* (swarna moola), *Sida cordifolia* (bala), *Curculigo orchioides* (kali musale), *Smilax chinensis*, *Dioscorea* spp., *Rubia cordifolia* (manjistha), *Sussurea costus* (kostam), *Rauvolfia serpentina* (sarpaganda), *Corallocarpus epigaeus* (akasa garudan), *Plumbago zeylanica* (chithrak mool) and *Vetiveria zizanioides*. Roots of dasamoola plants viz., *Desmodium gangeticum*, *Pseudarthria viscida*, *Tribulus terrestris*, *Solanum surattense* (kantakari), *Aegle marmelos* (bael), *Stereospermum personatum* (padiri), *Oroxylum indicum*, *Gmelina arborea* (shivani), *Clerodendrum phlomides* (thazuthalai), *Solanum anguivi* (mulsundai) are used frequently. *Hemidesmus indicus* (Nannari) is endemic to peninsular India and used in higher volumes both fresh for pickles and dried for medicine and a cool drink. *Pongamia pinnata* root is used to cure epilepsy and traditional healers near Palamaner (Andhra Pradesh) treat 1000 epilepsy cases. The roots of *Catharanthus roseus* (Nithyakalyani) are harvested and exported from Tuticorin port.

Withania somnifera cultivation became a normal practice in the northern states such as Madhya Pradesh and Rajasthan. Due to higher market demand and price, the roots of *Rauvolfia serpentina* and *Sussurea costus* are cultivated on commercial scale. In commercial manufacturing, only certified farmed materials can be used. Vetiver is grown on a large scale for commercial purposes. In the case of root drugs, the cultivation period is long and the crop maintenance costs are high, moisture loss and impurities are the major issues in obtaining quality material at a fixed price. As a result of this problem, finished product production costs are greater. For commercial production, the National Biodiversity Authority (NBA) insists on using grown medicinal plants. For the conservation of the flora, good collection practices must be followed, however, the aggressive collection done by the local people has resulted in a scarcity of raw herbs in their habitats. Several thousand tonnes of *Salacia* roots are currently being used for export and home purposes. When the riches are depleted, there will undoubtedly be a scarcity of roots. To fulfill future demands, the Forest Department and major landowners should cultivate *Salacia* on a wider scale. Using vegetative cuttings, *Plumbago zeylanica* can be grown at the edges of agricultural lands.

Asparagus curillus: Root powder is administered for dysuria, diabetes and dysentery.

Asparagus racemosus: Root powder administered orally as aphrodisiac, also used in stomach ache, urinary disorders, nervous disorders, inflammation, liver diseases, infectious diseases, ulcers and cancer.

Berberis aristata: Root extract used in eye diseases, diarrhoea, jaundice, skin diseases, syphilis, chronic rheumatism, urinary disorders, diabetes, jaundice and gastro intestinal problem.

Berberis chitria: Root extract is very useful for fever, jaundice, diabetes, stomach disorders, rheumatism, eye, ear and skin diseases.

Bergenia ciliata: Root powder is very useful in kidney stones and ulcers.

Boerhavia diffusa: Juice of fresh roots is used as eye drops; very useful tonic to liver problem, jaundice, asthma with antimicrobial activity.

Byttneria herbacea: Root paste is used in fever and leucorrhoea.

Carissa spinarum: It is used in rheumatism, having strong purgative property so used as ingredients in purgative preparations.

Drimia indica: Juice of the bulb is used in cough, bronchitis, nematode infection, pyrexia dropsy, respiratory ailment, bone joint complications, skin disorders, epilepsy and cancer.

Geranium wallichianum: Root extract is useful in backache, gout, bone strengthening, hepatitis, liver problems, and premature delivery.

Helicteres isora: Orally use of crushed roots are used in excessive appetite, empyema and stomachic and it is having antioxidant and anticancer properties.

Plumbago zeylanica: Paste of root and fruit is applied on the affected part for wound healing and also having anti-atherogenic, cardiotonic, hepatoprotective and neuroprotective.

b) Bark Drugs

Cinchona ledgeriana: The barks yield the drug quinine. The quinine and its related alkaloids are obtained from the bark of cinchona. Quinine is the most important alkaloid of the bark of cinchona. In the form of its salts, such as sulfate, bisulfate, hydrochloride and dihydrochloride, it is used for the prevention and treatment of malaria.

Soymida febrifuga: Rheumatoid arthritis, asthma and ulcers are all treated with bark extracts. The bitter resin in the bark decoction is used to treat vaginal infections, rheumatic problems and stomach pains. The bark is applied in wounds, dental disorders, uterine bleeding and haemorrhage as an anti-cancer treatment. It is used as an acrid, refrigerant, anthelmintic agent, aphrodisiac, laxative treats tridosha fevers, cough and asthma. It removes impurities from the blood and treats ulcers, leprosy and dysentery and it has anti-inflammatory properties. In advanced stages of dysentery and diarrhoea, the bark is used to treat intermittent fevers and general debility. It is similar to cinchona, has an effective antimalarial and antibacterial properties. The bark is astringent to the bowels and is used in Yunani medicine as a decoction for fevers.

Ficus religiosa: It is commonly known as peepal in India, belongs to the family Moraceae. It has been reported to be used in the traditional system of Ayurveda for the treatment of diabetes. It has been shown to possess a wide spectrum of *in vitro* and *in vivo* pharmacological activities: antidiabetic, hypolipidemic, anticonvulsant, anti-inflammatory, analgesic, antimicrobial, antiviral, antioxidant, antitumor, antiulcer, antianxiety, anthelmintic, antiasthmatic, immunomodulatory, estrogenic, endothelin receptor antagonist, apoptosis inducer, cognitive enhancer, and antihypertensive. Decoction prepared from the bark is used in the treatment of diabetes.

***Holarrhena antidysenterica*:** This plant, also known as kutaj is native to Asia and Africa's tropical and subtropical regions. It can be found abundantly in India, particularly in the Himalayan ranges. In India, it has traditional and folklore values.

Pterocarpus marsupium (Indian kino tree) belongs to the family Fabaceae. *Pterocarpus marsupium* contains terpenoids and phenolic compounds: β-sitosterol, lupenol, aurone glycosides, epicatechins and iso-flavonoids. *P. marsupium* is known for its antidiabetic activity.

Eugenia jambolana belongs to the family Myrtaceae. It is an evergreen tropical tree of 8 to 15 m in height. The bark is scaly gray and the trunk is forked. The bark of jamun is rich in several bioactive compounds including quercetin, betulinic acid, β- sitosterol, eugenin, ellagic and gallic acid, bergenin, tannins and flavonoids.

Boswellia serrata: Hot decoction of bark is used in wounds, chronic inflammatory diseases and diarrhoea.

Syzygium cumini: Bark extract is useful for diabetes, sore throat, bronchitis, asthma, thirst, biliousness, dysentery and ulcers.

Terminalia arjuna: Bark is very useful in the treatment of pneumonia, fractures, ulcers and hepatic and shows hypocholesterolaemia, antibacterial, antimicrobial, antitumoral, antioxidant, antiallergic and antifeedant, antifertility and anti-HIV activities.

c) Flower, Fruit and Seed drugs

Aegle marmelos: Astringent, anti-diarrheal, anti-dysentery, demulcent, antipyretic, laxative.

Aesculus indica: Rheumatism (Warm paste applies on affected part).

Amelanchier canadensis: The juice of the ripe fruit is used in the treatment of diarrhoea.

Butea monosperma: The flower extract with water is useful in gastritis. Gum from bark known as bengal kino is useful in treatment of ulcer, diarrhoea, anti-oxidative, anti-inflammatory, hepatoprotective and anti-cancer activities.

Callicarpa macrophylla: It is very helpful in urinary disorders, skin problem, diarrhoea, defecation and hair care.

Celastrus paniculatus: Boiled solution of fruits and seeds with oil used in piles, gout, rheumatism, cold, dysentery, diarrhoea, leprosy, snake bite, wound, anti-spasmodic, carminative and anthelmintic properties and also used for the treatment of epilepsy, mental ailments, chronic diarrhoea, dysentery, bronchial catarrh, intermittent fevers and tumours, pneumonia and leucorrhoea.

Cordia myxa: Fruit is very useful in common cold and cough.

Corylus colurna: Direct use of nut or its tonic is used as diuretic and aphrodisiac.

Cydonia oblonga: Seeds are used in diarrhoea, dysentery, cough, sore throat and bronchitis, intestinal colic and constipation and also used for allergic rhinitis and asthma.

Datura stramonium: Flower juice is used in ear ache having analgesic and antiasthmatic activities.

Dicliptera bupleuroides: Decoction of seeds with water is very useful in dysentery, respiratory and urinary tract infections, digestive disorders, sexual problems, skin diseases, liver ailments, wound healing, blood vomiting, malaria, piles, epilepsy, jaundice, gummosis, roundworms and hepatitis.

Ficus palmata: Fruits are beneficial in lung and bladder diseases. These are the good source of minerals, phosphorus and a small amount of vitamin C. The latex is useful in wound healing.

Glycosmis pentaphylla: Seeds of a plant is used in curing vomiting

Hygrophila auriculata: Root is used for rheumatism, inflammation, jaundice, hepatic obstruction and pain. The whole plant extract is used in the treatment of urinary infection, gout, hepatic obstruction and as a diuretic.

Prunus armeniaca: The fruit having high in carotene and vitamin C provides a valuable source of food. Amygdalin, a chemical extracted from apricot kernels is used as an alternative treatment for cancer which has antioxidant and antimicrobial properties.

Punica granatum: Fruit is very useful to worm infestation with antimicrobial, antioxidant, anti-inflammatory and anti-cancerous activities.

d) Leaf Drugs

Gymnema sylvestre: Boiled leaves with oil when applied externally 3-4 drops twice per day for eyes to reduce pain. Direct use of milky juice from crushed fresh leaves when applied once daily for 3 days also helpful in eye problems along with antidiabetic properties.

Artemisia annua: It is used throughout Asia and Africa as tea and press juice to treat malaria and related symptoms (fever, chills). Its active ingredient, artemisinin has been developed as an antimalarial drug and is used worldwide.

Ephedra gerardiana: It is a significant medicinal herb that belongs to the family Ephedraceae. This plant is mostly found in dry rocks and sandy soil as its growth requires high and dry mountain deserts. It is used to cure a variety of diseases such as asthma, nasal congestion, hay fever, inflammation or swelling, urinary disorders, cardiac disorders, cough, chills and cold.

Ageratum conyzoides: Leaf paste is very useful in wound healing and treatment of leprosy, diarrhea, dysentery, intestinal colic, rheumatism and fever.

Anaphalis triplinervis: It is used in wounds, skin diseases, flu, fever, nausea and eye infections.

Bombax ceiba: Leaf paste applies over wounds with good property of pain killer and improves gynaecological disorders.

Cleome viscosa:Leaf juice is very useful for ear diseases, anthelmintic, antiseptic, carminative, antiscorbutic, sudorific, febrifuge and cardiac problems.

Commelina benghalensis: Paste of leaves is used for curing swelling due to honeybee biting.

***Anisomeles indica*:** Fresh leaves as well as greenish parts used in the treatment of cough and cold, chronic rheumatism, psoriasis and snake bites.

Buchanania cochinchinensis: It is very effective in dental pain when gargle the juice twice daily for asthma and cancer.

Calotropis gigantea: Warmed fresh leaves are used for the wound, tooth ache, ear ache, epilepsy, diarrhoea and mental disorders.

Cydonia oblonga: Leaves having phenolic compounds as antioxidant properties, vitamin E, carotenoids, L-ascorbic acid and other organic acids in folk medicine for their sedative, antipyretic, anti-diarrheic and antitussive properties along with treatment of various skin diseases.

Datura metel: Boiled and warm leaves are used in the affected part for antibiotics, antibacterial and antimicrobial activity.

Ficus auriculata: It is used for gastrointestinal problems using 50-100ml fresh leaves juice regularly upto 10 days.

Grevillea robusta: Paste of fresh leaves with vegetable ghee and coconut oil is very useful for burning and burning sensation.

Hydrocotyle sibthorpioides: Leaf juice is administered orally for mental disorder, also used in various problems such as dysentery, diarrhoea, piles, rheumatism, digestive, diuretic, vermifuge and menstrual problem.

Justicia adhatoda: Juice of leaves and flowers is used for cough, fever, coryza, respiratory problems.

***Kalanchoe pinnata*:** Paste of leaves is used for wound healing.

Extraction of Plant Drugs

Medicinal plants are extracted and processed for direct consumption as herbal or traditional medicine. The concept of preparation of medicinal plant involves the proper and timely collection of the plant, authentication by an expert, adequate drying and grinding. This is followed by extraction, fractionation and isolation of the bioactive compounds where applicable. In addition, it comprises determination of quantity and quality of bioactive compounds. Extraction of medicinal plants is a process of separating active plant materials or secondary metabolites such as alkaloids, flavonoids, terpenes, saponins, steroids and glycosides from inert or inactive material using an appropriate solvent and standard extraction procedure. Plant materials with high content of phenolic compounds and flavonoids are found to possess antioxidant properties, and hence are used to treat age related diseases such as alzheimer's disease, parkinsonism, anxiety and depression.

Several methods are used in the extraction of medicinal plants such as maceration, infusion, decoction, percolation, digestion and soxhlet extraction, superficial extraction, ultrasound-assisted and microwave-assisted extraction. In addition, thin-layer chromatography (TLC), high-performance liquid chromatography (HPLC), paper chromatography (PC) and gas chromatography (GC) are used in separation and purification of the secondary metabolites. The choice of an appropriate extraction method depends on the nature of the plant material, solvent used, pH of the solvent, temperature and solvent to sample ration. It also depends on the intended use of the final products.

Solvent Extraction

The solvent used for the extraction of medicinal plants is also known as the menstruum. The choice of solvent depends on the type of plant, part of plant to be extracted, nature of the bioactive compounds and the availability of solvent. In general, polar solvents such as water, methanol and ethanol are used in extraction of polar compound whereas nonpolar solvents such as hexane and dichloromethane are used in extraction of nonpolar compounds. During liquid-liquid extraction, the conventional way is to select two miscible solvents such as water-dichloromethane, water-ether and water-hexane. In all the combinations, water is present because of its high polarity and miscibility with organic solvent. The compound to be extracted using liquid-liquid extraction should be soluble in organic solvent but not in water to ease separation. Furthermore, solvent used in extraction is classified according to their polarity, from *n*-hexane which is the least polar to water the most polar.eg:*n*-hexane, petroleum ether, diethyl ether, ethyl acetate, chloroform, dichloromethane, acetone, *n*-butanol, ethanol, methanol and water.

Properties of Solvent Extraction

i) Water: It is the most polar solvent and is used in the extraction of a wide range of polar compounds.

Advantages: It dissolves a wide range of substances. It is cheap, nontoxic, nonflammable and highly polar.

Disadvantages: It promotes bacterial and mold growth. It may cause hydrolysis and a large amount of heat is required to concentrate the extract.

ii) Alcohol: It is also polar in nature, miscible with water and could extract polar secondary metabolites.

Advantages: It is self-preservative at a concentration above 20%. It is nontoxic at low concentration and a small amount of heat is required for concentrating the extract.

Disadvantages: It does not dissolve fats, gums and wax. It is flammable and volatile.

iii) Chloroform: It is a nonpolar solvent and is useful in the extraction of compounds such as terpenoids, flavonoids, fats and oils.

Advantages: It is colorless, has a sweet smell and is soluble in alcohols. It is also well absorbed and metabolized in the body.

Disadvantages: It has sedative and carcinogenic property.

iv) Ether: It is a nonpolar solvent and is useful in the extraction of compounds such as alkaloids, terpenoids, coumarins and fatty acids.

Advantages: It is miscible with water, has low boiling point and is tasteless in nature. It is also a very stable compound and does not react with acids, bases and metals.

Disadvantages: It is highly volatile and flammable in nature.

v) Ionic liquid (green solvent)**:** This is a unique solvent of extraction and is highly polar and extremely heat stable. It can remain in a liquid state even at 3000°C and usable where high temperature is applicable. It has extreme miscibility with water and other solvent and is very suitable in the extraction of polar compounds.

Advantages: It is an excellent solvent that attracts and transmit microwave and hence it is suitable for microwave-assisted extraction. It is nonflammable and is useful for liquid-liquid extraction and highly polar.

Disadvantage: It is not ideal for preparation of tinctures.

The following factors are to be considered in choosing extraction method

Stability to heat: Heat stable plant material is extracted using soxhlet extraction or microwave-assisted extraction, whereas plant materials that are not heat stable are extracted using maceration or percolation.

Nature of solvent: If the solvent of extraction is water, maceration is a suitable method but for volatile solvent, percolation and soxhlet extraction are more appropriate.

Cost of the drug: Cheap drugs are extracted using maceration, whereas costly drugs are preferably extracted using percolation.

Duration of extraction: Maceration is suitable for plant material requiring long exposure to the menstruum, whereas techniques such as microwave- or ultrasound-assisted extraction is used for a shorter duration.

Final volume required: Large volume products such as tinctures are prepared by maceration, whereas concentrated products are produced by percolation or soxhlet extraction.

Intended use: Extracts intended for consumption by human are usually prepared by maceration whereas products intended for experimental testing are prepared using other methods in addition to maceration.

Methods of Extraction

i) Maceration

This is an extraction procedure in which coarsely powdered drug material, either leaves or stem bark or root bark, is placed inside a container; the menstruum is poured on top until completely covered with the drug material. The container is then closed and kept for at least three days. The content is stirred periodically and if placed inside bottle, it should be shaken time to time to ensure complete extraction. At the end of extraction, the micelle is separated from marc by filtration or decantation. Subsequently, the micelle is then separated from the menstruum by evaporation in an oven or on top of water bath. This method is convenient and very suitable for thermolabile plant material.

ii) Infusion

This is an extraction process such as maceration. The drug material is ground into fine powder and then placed inside a clean container. The extraction solvent hot or cold is then poured on top of the drug material, soaked and kept for a short period of time. This method is suitable for extraction of bioactive constituents that are readily soluble. In addition, it is an appropriate method for preparation of fresh extract before use. The solvent to sample ratio is usually 4:1 or 16:1 depending on the intended use.

iii) Digestion

This is an extraction method that involves the use of moderate heat during extraction process. The solvent of extraction is poured into a clean container followed by powdered drug material. The mixture is placed over water bath or in an oven at a temperature about 50° C. Heat is applied throughout the extraction process to decrease the viscosity of extraction solvent and enhance the removal of secondary metabolites. This method is suitable for plant materials that are readily soluble.

iv) Decoction

This is the process that involves continuous hot extraction using specified volume of water as a solvent. A dried, ground and powdered plant material is placed into a clean container. Water is poured and stirred. Heat is then applied throughout the process to hasten the extraction. The process is lasted for a short duration usually about 15min. The ratio of solvent to crude drug is usually 4:1 or 16:1. It is used for extraction of water soluble and heat stable plant material.

v) Percolation

The apparatus used in this process is called percolator. It is a narrow cone shaped glass vessel with opening at both ends. A dried, ground and finely powdered plant material is moistened with the solvent of extraction in a clean container. More quantity of solvent is added and the mixture is kept for a period of 4h. Subsequently, the content is then transferred into percolator with the lower end closed and allow to stand for a period of 24h. The solvent of extraction is then poured from the top until the drug material is completely saturated. The lower part of the percolator is then opened and the liquid allowed to drip slowly. Some quantity of solvent is added continuously and the extraction taken place by gravitational force, pushing the solvent through the drug material downward. The addition of solvent stopped when the volume of solvent added reached 75% of the intended quantity of the entire preparations. The extract is separated by filtration followed by decantation. The marc is then expressed and final amount of solvent is added to get required volume.

vi) Soxhlet extraction

This process is otherwise known as continuous hot extraction. The apparatus is called soxhlet extractor made up of glass. It consists of a round bottom flask, extraction chamber, siphon tube and condenser at the top. A dried, ground and finely powdered plant material is placed inside porous bag (thimble) made up of a clean cloth or strong filter paper and tightly closed. The extraction solvent is poured into the bottom flask, followed by the thimble into the extraction chamber. The solvent is then heated from the bottom flask, evaporates and passes through the condenser where it condenses and flow down to the extraction chamber and extract the drug by coming in contact. Consequently, when the level of solvent in the extraction chamber reaches the top of the siphon, the solvent and the extracted plant material flow back to the flask. The entire process continues repeatedly until the drug is completely extracted, a point when a solvent flowing from extraction chamber does not leave any residue behind. This method is suitable for plant material that is partially soluble in the chosen solvent and for plant materials with insoluble impurities. However, it is not a suitable method for thermolabile plant materials.

Advantages: Large amount of drug can be extracted with smaller amount of solvent. It is also applicable to plant materials that are heat stable. No filtration is required and high amount of heat could be applied.

Disadvantages: Regular shaking is not possible and the method is not suitable for thermolabile materials.

vii) Microwave-assisted extraction

This is one of the advanced extraction procedure in preparation of medicinal plants. The technique uses mechanism of dipole rotation and ionic transfer by displacement of charged ions present in the solvent and drug material. This method is suitable for extraction of flavonoids. It involves the application of electromagnetic radiation in frequencies between 300 MHz and wavelength between 1cm and 1 m. The microwaves applied at frequency of 2450 Hz yield energy between 600 and 700 W. The technique uses microwave radiation to bombard an object, which can absorb electromagnetic energy and convert it into heat. Subsequently, the heat produced facilitates movemcnt of solvent into the drug matrix. When polar solvent is used, dipole rotation and migration of ions occur, increase solvent penetration and assist extraction process. However, when nonpolar solvent is used, the microwave radiation released will produce only small heat. Hence, this method does not favor use of nonpolar solvents.

Advantages: Microwave-assisted extraction has special advantages such as minimizing solvent and time of extraction as well as increase in the outcome.

Disadvantages: This method is suitable only for phenolic compounds and flavonoids. Compounds such as tannins and anthocyanins may be degraded because of high temperature involved.

viii) Ultrasound assisted extraction

This process involves the application of sound energy at a very high frequency greater than 20 KHz to disrupt plant cell and increase the drug surface area for solvent penetration. Consequently, secondary metabolites will be released. In this method, plant material should dry first, ground into fine power and sieved properly. The prepared sample is then mixed with appropriate solvent of extraction and packed into the ultrasonic extractor. The high sound energy applies hasten the extraction process by reducing the heat requirements.

Advantages: Ultrasound assisted extraction is applicable to small sample. It reduces the time of extraction and amount of solvent used and maximizes the yield.

Disadvantages: This method is difficult to be reproduced and also high amount of energy applied may degrade the phytochemicals by producing free radical.

Fractionation and Purification Methods

Fractionation is a process of separation of plant extracts into various fractions. It further segregates the fractions into portions comprising a number of compounds. The process continues until pure compound is isolated. When several solvents are required for the fractionation, they should be added according to the order of increasing polarity. Fractionation techniques are basically classified into physical and chemical method.

a) Chemical Method

This extraction method is based on the type of functional groups possessed by a compound in the given mixture. Separation or purification can be achieved by chemical reactions using appropriate reagents.

b) Physical Method

Physical methods used in separation of compounds from mixtures include separation funnel method, chromatographic technique, fractional distillation, fractional crystallization, fractional liberation and sublimation. These separations are briefly explained here below.

i) Separation funnel method: When four different solvents (*n*-hexane, chloroform, acetone, and *n*-butanol) are selected, fractionation begins by moistening or complete dissolution of crude extract with 250ml of water. This is followed by transfer into a separating funnel, shaken and allowed to settle. Furthermore, to 250ml of *n*-hexane, the least polar solvent is added and shaken. The content can settle and the bottom of the separating funnel is opened to remove the aqueous layer. The remaining content in the separating funnel is poured into a clean container to get *n*-hexane fraction. Equal volume of *n*-hexane is added again, shaken and separated. The addition continued until after adding *n*-hexane and shaken no reasonable quantity of extract appeared to move into the *n*-hexane portion. Similar cycle is performed for chloroform, acetone, *n*-butanol to get chloroform, acetone and *n*-butanol fractions. The remaining portion left after the fractionation is termed as residual aqueous fraction (RAF) as the crude extract is first dissolved in water.

ii) Fractional distillation: This is the process of separating or purifying compounds from a mixture. It is usually used in separation of hydrocarbons such as crude oil, citral and eucalyptol. Purification is achieved based on the differences in their boiling points. Fractional distillation apparatus is constructed in such a manner that when heat is applied each compound will evaporate and separates at its boiling point. Consequently, each compound fractionated will condense and collected as a separate entity through several siphons attached to fractional distillation apparatus.

iii) Fractional crystallization: Large numbers of compounds that exist naturally in plant extracts are crystal in nature. Separation is achieved via formation of crystals during concentration of an extract using heat or refrigeration.

iv) Fractional liberation: This method is suitable for separating compounds that can easily form precipitate from the mixture. The precipitate is usually formed by changing the compounds into their salt form. Fractional liberation is commonly applicable in purification of cinnamon alkaloids.

v) Sublimation: This method involves changing from solid to gaseous state without passing through liquid state. Substances such as camphor and volatile oils when heated get separated and converted directly into gas.

vi) Chromatographic techniques: These are special techniques used in separation of compounds from mixtures based on their size, shape and charge. The concept of chromatography involves the use of mobile phase, which is the solvent of extraction and the stationary phase such as silca gel and sephadex mixed with a calcium sulfate as a binder. Silica gel is used for parting amino acids, sugars, fatty acids, lipids and alkaloids. Sephadex is applicable in isolation of proteins and amino acids. Aluminum is useful in separation of vitamins, carotenes, phenols, steroids and alkaloids. Cellulose powder is used in separation of amino acids, food dyes and alkaloids. Celite is applicable in separation of organic cations and steroids. Various mechanisms are involved in separation of compounds using chromatographic techniques viz., adsorption, partition, affinity, ion exchange or size exclusion. Chromatographic techniques include paper chromatography, thin layer chromatography, column chromatography, liquid chromatography, gas chromatography and high performance liquid chromatography.

Types of value addition

The existing system of medicinal plant collection or cultivation in India needs substantial improvements for the harnessing the full potential of this important industry. Improvements are needed especially in the areas of post collection handling, processing and product presentation. The following direct or indirect value addition techniques shall be followed without much capital to increase the value of medicinal plant materials and reduce the contamination and wastages.

a) Direct value addition

- **Collection at right season and right maturity:** To get the right composition of phytochemicals or essential oils, collection or harvesting should be carried out at appropriate stages and seasons.
- **Grading and sorting:** Basic procedures like grading and sorting are the effective means of getting higher value for the produce with minimal efforts.
- **Cleaning:** Cleaning technique should be followed to remove unwanted foreign materials and pollutants.
- **Proper drying:** Depending upon the needs, the drying has to be carried out to retain the desired quality and to avoid losses.
- **Packaging:** Keeps contents from extraneous matter or from loss of the material under normal conditions of handling shipment or storage. Different categories of the plant materials require different packaging considerations.

b) Indirect value addition

- **Organic certification**: Of more than 400 plants species used for production of medicine by the Indian Herbal Industry, fewer than 20 species are currently under cultivation in different parts of the country. Organic cultivation of selected medicinal and aromatic plants and certification can be profitable to cultivators with buy back arrangement with buyers or user industries.

c) Semi-processing of the medicinal plants to value added products

Semi processing like producing powder, tablets and extracts of herbal materials does not require very expensive processing machineries and can be carried for many of the herbs available locally based on the demand.

Value added products from Medicinal Plants

- **Herbal tea:** Dried herbal plant parts powdered to specific mesh size and packed in attractive tea bags or sachets are becoming popular world over. China is credited with utilizing more than 200 plants as herbal tea.
- **Herbal juice, sherbet, herbal drink:** These are traditionally used in several countries and are currently aggressively marketed by countries like China, Thailand, etc. Herbal beer is being marketed by a firm in India. Herbal drinks

such as Nannari (*Decalepis hamiltonii*), Sugandhipala (*Hemidesmus indicus*), Aloe (*Aloe vera*), Noni (*Morinda citrifolia*), etc. are popular products.

- **Herbal chocolates and sweets**: Sugar free sweets, diet chocolates, toffees, candies, etc., are being marketed internationally. Use of herbal extracts in food products is not a new idea. Solvent extracts of several medicinal plants with specified minimum concentration of chemicals are traded globally. A number of small, medium and large industries are currently producing hundreds of such extracts.
- **Extraction of medicinal chemicals:** Technologies are available for isolation of pure phytochemicals such as alkaloids, glycosides, flavonoids, steroids, coumarins, etc. from several medicinal plants and these are regularly used by the drug industry.
- **Plant drugs:** This is an age old well known industry manufacturing plant based drugs in traditional systems of medicine such as ayurveda, siddha, homeopathy, etc. and in modern medicine.
- **Herbal cosmetics/cosmeceuticals:** Cosmeceuticals are a blend of cosmetics and pharmaceuticals. For example, a dandruff shampoo is a shampoo meant to be used for personal hygiene. However, it is also used to treat a medical condition, giving it properties of a pharmaceutical. Because it has this dual purpose, it is known as a cosmeceutical. Though herbs have been used since several centuries as beauty aids, products made from medicinal plants in combination with other beauty enhancing products are becoming popular internationally.
- **Nutraceuticals:** Nutraceutical (nutrition + pharmaceutical) refers to extracts of foods claimed to have a medicinal effect on human health. The nutraceutical is usually contained in a medicinal form such as capsule, tabiet or powder in a prescribed dose. More rigorously, nutraceutical implies that the extract or food is demonstrated to have a physiological benefit or provide protection against a chronic disease. This is a sunrise industry with a large potential for small and medium enterprises.
- **Functional foods:** Functional foods are defined as being consumed as part of a usual diet but are demonstrated to have physiological benefits and/or reduce the risk of chronic disease beyond basic nutritional functions e.g. diabetic flour.
- **Herbal pesticides:** This enterprise has enormous potential in future.
- **By-product utilization:** Many enterprises can be established at rural level making use of by products/ waste material to produce value added products.

Suggested Readings

Marla R Emery and Rebecca J Mclain 2001 *Non-Timber Forest Products: Medicinal Herbs, Fungi, Edible Fruits and Nuts, and Other Natural Products from the Forest*, Boca Raton, Florida: CRC Press

Rai MK Geoffrey A Cordell Jose L Martinez Mariela Marinoff and Luca Rastrelli 2012 *Medicinal Plants: Biodiversity and Drugs*, New York: CRC Press

Rajeev Joshi and Ramesh Chhetri 2021 *A Textbook of Non-Timber Forest Products,* Bhotahity, Kathmandu Nepal: Heritage Publishers and Distributors Pvt. Ltd

Savitha P Ushakumari R and Manivannan A 2022 *Indian Medicinal Trees,* Delhi: Narendra Publishing House,

Sophia Grieve 2012 *Medicinal Plants, Trees and Herbs,* Singapore: Star Rising Publishers

CHAPTER 10
Spices and Edible Fruits

M.P. Divya and R. Ravi
Tamil Nadu Agricultural University, Coimbatore

SPICES

Spices are natural plant or vegetable products and mixtures that are used whole or ground to improve the flavour, fragrance and piquancy of food. Spices are aromatic, vegetable based compounds that are used in cooking to enhance the flavour and pungency of food. Spices provide flavour, aroma and colour to food, making it more attractive to eat. They have a strong odour as well as a sweet or bitter flavour. "Spices are plant parts such as seed, bark, berries, buds, floral parts, fruits, kernels, leaves, rhizome, latex and so on. Instead of nutrition, these aromatic vegetable compounds are utilized to season meals in whole, broken, or ground form. A wide diversity of spices, except cardamom, ginger, zedoary and pepper which can be found naturally in forests are cultivated crops. The following spices *viz.,* cardamom, pepper, turmeric, cinnamon, nutmeg, cloves and coriander are found in the forests.

Major Spices

i) Cardamom

There are two well known species yielding the cardamoms, the more important is *Elettaria cardamomum,* and the other species is *Amomum subulatum*. Cardamom, popularly known as the "Queen of Spices," has a history, dating back to the *Vedic* period. *Elettaria cardamomum*, commonly known as green or true cardamom is an herbaceous, perennial plant in the ginger family, native to southern India. It is the most common of the species whose seeds are used as a spice called cardamom. True cardamom may have been used in Ayurveda medicine as early as the 4th century BC. Ground cardamom is an ingredient in many Indian curries and is a primary contributor to the flavour of masala chai. In Iran and India, cardamom is used to flavour coffee and tea. In addition to its native range, it is grown in Nepal, Vietnam, Cambodia, Thailand, Sri Lanka and Central America. In India, the states of Sikkim and Kerala are the main producers of cardamom. They rank highest both in cultivated area and in production.

ii) Pepper

Various species of pepper plants are found in India, the commonest being *Piper longum,* found in the damp evergreen forests of West Bengal, Assam and the West Coast. Black pepper (*Piper nigrum*) is a flowering vine in the family Piperaceae, which is usually dried and used as a spice and seasoning. The fruit is a drupe (stone fruit) which is about 5 mm in diameter (fresh and fully mature), dark red and contains a stone which encloses a single pepper seed. Peppercorns and the ground pepper derived from them may be described simply as *pepper* or more precisely as black pepper (cooked and dried unripe fruit), green pepper (dried unripe fruit), or white pepper (ripe fruit seeds). Black pepper is native to the Malabar Coast of India and the Malabar pepper is extensively cultivated in these areas and in other tropical regions also.

iii) Turmeric

The roots of turmeric (*Curcuma aromatica*), a common forest plant of the South Asian region, predominantly in the eastern Himalayas and especially in South India. It is native to Tamil Nadu and is used as a substitute for true turmeric, which is used as an ingredient in curries and colouring the foodstuff. The wild turmeric is one among 80 members of the plant family Zingiberaceae. The perennial foliage dies down in late autumn and the rhizomes remain dormant in winter. The inflorescence appears in early spring from the base of the rhizomes. During the summer monsoon season and the immediately following weeks, the plant grows fast and vigorously. The stalk grows to about 20-30 cm tall and is crowned with enlarged coloured bracts with pink tips. Leaves often appear even after the flowers. When in full growth, the plants can reach a height of about 40 cm (16 in) tall.

iv) Cinnamon

Cinnamomum verum also called Ceylon cinnamon, an evergreen tree of the Lauraceae family and the spice derived from its bark. Cinnamon is native to Sri Lanka, the neighbouring Malabar Coast of India, and Myanmar (Burma) and is also cultivated in South America and the West Indies. The spice consists of the dried inner bark is brown in colour and has a delicately fragrant aroma and a warm sweet flavour. Cinnamon is used to flavour a variety of foods viz., confections, curries and beverages and bakery goods.

v) Nutmeg

Myristica fragrans, a tropical evergreen tree of the Myristicaceae family and the spice made from its seed. It is made by grinding the seed of the fragrant nutmeg tree into powder. The spice has a distinctive pungent fragrance and a warm, slightly sweet taste. It is used to flavour many kinds of baked goods, confections, puddings, potatoes, meats, sausages and sauces, vegetables and such beverages as eggnog. The seeds are dried gradually in the sun over six to eight weeks. During this time the nutmeg shrinks away from its hard seed coat until the kernels rattle in their shells when shaken. The shell is then broken with a wooden club and the nutmegs are picked out. Dried nutmegs are greyish brown ovals with furrowed surfaces.

vi) Cloves

Cloves (*Syzygium aromaticum*) are the aromatic flower buds belongs to the family Myrtaceae. It is native to Maluku Islands (or the Moluccas) in Indonesia and are commonly used as a spice. The clove tree is an evergreen that grows up to 8-12 m (26-39 ft) tall, with large leaves and crimson flowers grouped in terminal clusters. It can grow in tropical and subtropical areas. The flower buds initially have a pale hue, gradually turn green and then transition to a bright red when ready for harvest. Cloves are harvested at 1.5-2 cm long and consist of a long calyx that terminates in four spreading sepals and four unopened petals that form a small central ball.

vii) Coriander

Coriander grows wild over a wide area of Western Asia and Southern Europe. It is hard to define exactly where this plant is wild and where it only recently established itself. It is a soft plant growing to 50 cm (20 in) tall. The leaves are variable in shape, broadly lobed at the base of the plant, slender and feathery higher on the flowering stems. All parts of the plant are edible, but the fresh leaves and the dried seeds are the parts most traditionally used as a spice in cooking.

Value added products

a) Ready made spice powder and paste

Powdered spice in air tight packaging material is of enormous demand. Increasing urbanization paired with a rise in number of working women has reduced the time of cooking. Consequently, homemakers have started demanding readymade spice powder that includes chilli powder, cumin powder, fennel powder, black pepper powder, turmeric powder. Also popular are ready made paste of onion, garlic, ginger in packet form. An official report from Everest Spices Ltd. reveals their exports about 10 per cent of its products to the US, West Asia, Singapore, Australia, New Zealand and East Africa. The total market size of branded spices is estimated at 6,600 crore and is growing at 14 per cent annually.

b) Spices extractives

i) Essential oil

Essential oils are major flavouring constituents of spices, highly concentrated about 75-100 times than the fresh spice.

Coriander: Major component of volatile oil is linalool (67.7%) followed by alphapinene (10.5%), amma-terpinene (9.0%), geranyl acetate (4.0%), camphor (3.0%) and geraniol (1.9%). Minor components include beta-pinene, camphene, myrcene, limonene, p-cymene, borneoletc. Indian coriander oil differs from European oil in possessing a lower linalool content and comparatively higher linalyl acetate content. In cumin, the main constituent is cuminaldehyde and three other aldehydes up to 70%. Dominant monoterpenes hydrocarbons (total about 50-55%) are B-pinene, g-terpinene and p-cymene, plus myrcene and b- phellandrene and limonene with minor amount of sesquiterpenes hydrocarbons. Cumin oil is sometimes adulterated with synthetic cuminaldehyde, which is difficult to detect. The oil is a raw material for the production of thymol.

Fennel: Fennel seed oil, usually traded as fennel oil, is mainly obtained by steam distilling whole or crushed fruit with a yield of 1.5-6.5% and more recently by supercritical carbon dioxide extraction. In general, oil content is greatest in European and lowest in Asian varieties. The main constituents are transanethole (60-65%, up to 90%), fenchone (2-20%) estragole (methyl chavicol), limonene, camphene, a-pinene and other monoterpenes, fenchyl alcohol and anisaldehyde. Oil produced in Nigeria from fennel of Indian origin had 80% anethole content.

Fenugreek: Major constituents are the dihydroactinidiolide, 2,3-dihydrobenzofuran and 1-heanol totaling 7-9% with 20 other constituents at less than 3% and the remainder below 1%. The furanone derivative, sotolon, is reportedly mainly responsible for the characteristic fenugreek odour.

ii) Oleoresin

Oleoresin represents the complete flavour and non-volatile resinous fraction present in the spices. The aroma and taste fractions are proportionally blended to constitute the true essence of the natural spice. The oleoresin can be obtained in a single step by elimination of the steam distillation process.

iii) Derivatives of essential oil and oleoresins

It includes plated encapsulated forms of spice extractives, seasonings in dry carrier such as dextrose, salt or rusk powder. They impart the strength of good quality freshly ground spices and can be easily incorporated in the food.

iv) Dried whole spice

Dry ginger and kalaunji prices increased by Rs. 200 each to finish at Rs.15,200 to 18,200 /- and Rs 9,700-10,200 per quintal respectively.

Value added products from major spices

i) Black pepper

A variety of products are being made from black pepper. This also includes green pepper based products, black pepper and white pepper based products and pepper by-products. The major green pepper based products are canned green pepper, green pepper in brine, bulk packaged green pepper in brine, cured green pepper, frozen green pepper, freeze dried green pepper, dehydrated green pepper, green pepper pickle, mixed green pepper pickle, green pepper sauce and green pepper-flavoured products. Black pepper and white pepper based products include whole black pepper, sterilized black pepper, ground black pepper, cryo-ground black pepper powder, pepper oil and oleoresin, white pepper and white pepper powder. Other miscellaneous products from pepper are pepper-flavoured products, pepper extract, curry powder spice blends, peppersal, pepper mayonnaise, pepper cookies, etc.

ii) Cardamom

Major products of cardamom are bleached cardamom, decorticated seeds and seed powder, cardamom volatile oil and cardamom oleoresin, encapsulated cardamom, cardamom tea, cardamom coffee and cardamom soft drink mix.

iii) Ginger

Ginger powder, ginger oil, ginger oleoresins, encapsulated ginger, ginger preserves and salted ginger are the value added products from ginger.

iv) Turmeric

Major value added products are ground turmeric, turmeric oil, turmeric oleoresin and curcuminoids.

v) Chilli

Oleoresin is the main value added product that can be obtained from chilli. Other products from chillies are dehydrated chilli, canned chilli, brined/pickled chilli and fermented chilli.

vi) Nutmeg and mace

Nutmeg oil and mace oil, nutmeg oleoresin, nutmeg butter are the main value added products. By utilizing nutmeg pericarp (rind), many value added products have been developed viz., nutmeg (rind) pickle, nutmeg (rind) preserve from slices and nutmeg (rind) preserve from shreds, nutmeg (rind) candy, nutmeg (rind) sweet chutney and nutmeg (rind) powder.

vii) Cinnamon

In addition to cinnamon bark, various other products are obtained from the tree viz., bark oil, leaf oil, bark oleoresin, etc.

viii) Seed spices

Value added products such as ground spices, spices extractives, curry powder, consumer packed spices and organic spices can be obtained from seed spices.

ix) Curry powders / blends and mixes

Curry powder is an indigenous seasoning made from various spices (coriander, cumin, fennel, fenugreek) constitute the raw materials used in quality curry powder. The ingredients of curry will change according to different needs. The colour, form and taste of various curries are in accordance with the custom of various nations and regions. Consumers all over the world demand different curry powder. The export trade in curry powder at present is dominated by India. Curry powder is made from a blend of several spices, the number vary from a minimum of 5 to more than 20 depending on end uses.

EDIBLE FRUITS

Among the most extensively utilised non-timber forest products (NTFPs), wild edible fruits (WEFs) are major sources of nutrition, medicine and revenue for their users. Wild Edible Fruits (WEFs) have the potential to fight against poverty and starvation and can help promote dietary diversification and improve the rural economy as well. The potential of WEFs for contributing to food security, income generation and environmental services are under exploited. Though they are not significant in terms of global production and consumption systems but are rich in macro and micronutrients, so capable of combating micronutrients deficiencies "hidden hunger. They can deliver many health benefits apart from fulfilling physiological needs, thus conferred the status of functional foods. As such, WEFs can contribute as nutritional support for health improvement and help in development of various value added local products, which can help up-lift the socio-economic status of the local inhabitants.WEF species can produce fibre, fuel and a variety of processed products in addition to food (Table 10.1).

Table 10.1 Edible fruits and its uses

Scientific Name	Edible part	Mode of consumption and use	Edible Fruit
Aegle marmelos	Fruits Pulp & Unripe Fruits	Sherbet is made from mucilaginous pulp. Ayurvedic medicine from dried slices of immature fruits	
Alangium salvifolium	Fleshy Aril of Ripe Fruit	Tribals consume the arils of ripe fruits	
Allophylus serratus	Ripe Fruits	Fruits are edible and are employed in religious rites	
Antidesma acidum	Ripe Fruits	Tribals eat fully ripe acidic fruits, which are also prepared into pickles.	
Artocarpus lacucha	Ripe Fruits	Ripe fruits are eaten; they are utilised as vegetables, pickles and jams	

Scientific Name	Edible part	Mode of consumption and use	Edible Fruit
Bauhinia vahlii	Seed	Raw or boiled or roasted seed can be eaken	
Bridelia retusa	Ripe Fruits	Tribals prefer fully ripe fruits; unripe fruits are too acidic	
Buchanania lanzan	Ripe Fruit & Seed Kernel	Seed kernels are used as a confectionery and sold in the market when fruits are ripen	
Capparis sepiaria	Ripe Fruits	Ripe fruits are eaten by tribal children	
Capparis zeylanica	Ripe & Unripe Fruit	Mature fruits are cut into pieces and sun dried before being used in curries; ripe fruits are edible	
Carissa spinarum	Ripe Fruits	Fully ripe fruits are delicious whether eaten fresh or preserved in pickles; mature fruits have milky latex.	

Scientific Name	Edible part	Mode of consumption and use	Edible Fruit
Citrus medica	Ripe & Unripe Fruit	Unripe fruits are used as a vegetable or as a food ingredient; ripe acidic fruits are edible	
Citrus sinensis	Ripe & Unripe Fruit	Ripe fruits are consumed fresh, whereas immature fruits are preserved in pickles	
Coccinia grandis	Ripe & Unripe Fruit	Children eat immature fruits, adults eat mature fruits as vegetables and tribals consume ripe fruits on occasion	
Cordia dichotoma	Ripe Fruits	Ripe fruits are consumed despite being mucilaginous; mucilage from mature fruits is used as gum	
Dillenia pentagyna	Ripe & Unripe Fruit	Unripe fruits are eaten raw, but ripe fruits are utilised as a vegetable.	
Diospyros malabarica	Ripe Fruits	Tribals consume ripe fruits, despite their foul odour, when food is scarce	

Scientific Name	Edible part	Mode of consumption and use	Edible Fruit
Diospyros melanoxylon	Ripe Fruits	All tribal communities relish ripe fruits; seeds of young fruits are used as vegetables and fruits are offered in religious ceremonies	
Erycibe paniculata	Ripe Fruits	Forest inhabitants consume sweet and flavorful ripe fruits	
Ficus benghalensis	Ripe Fruits	Ripe fruits are consumed on a limited scale; birds eat the fruits in large quantities	
Ficus hispida	Ripe & Unripe Fruit	Young figs are eaten as a vegetable and ripe figs are consumed	
Ficus racemosa	Ripe & Unripe Fruit	Young figs are consumed as a vegetable; ripe fruits are consumed as a fruit	

Scientific Name	Edible part	Mode of consumption and use	Edible Fruit
Ficus religiosa	Ripe Fruit	Ripe and insipid fruits are consumed as a famine food	
Ficus semicordata	Ripe & Unripe Fruit	Immature fruits are frequently prepared as a vegetable whereas ripe fruits are occasionally consumed raw	
Flacourtia indica	Ripe Fruit	Tasty and acidic ripe fruits are favourite for children and adult tribals	
Gardenia gummifera	Pulp of Ripe Fruit	Pinkish pulp of ripe and mature fruits are edible	
Grewia asiatica	Ripe Fruit	Ripe acidic fruits are taken as such or as a refreshing drink	

Scientific Name	Edible part	Mode of consumption and use	Edible Fruit
Grewia hirsuta	Ripe Fruit	Ripe fruits are occasionally eaten by tribal communities	
Grewia rothii	Ripe Fruit	Fully ripe fruits contain inadequate pulp but are occasionally consumed by children	
Grewia tiliifolia	Ripe Fruit	Ripe fruits are taken by tribal communities	
Lepisanthes tetraphylla	Fleshy Aril of Ripe Fruit	White coloured arils of ripe fruits are often consumed	
Limonia acidissima	Pulp of Ripe Fruits	The fruit pulp is widely used by tribals as such or as chutney, drink, medicine and tribal rituals	
Litsea glutinosa	Ripe Fruit	Fruits, when fully ripe are eaten	

Scientific Name	Edible part	Mode of consumption and use	Edible Fruit
Madhuca indica (Mahua)	Mature Fruit & Flower	Berries are eaten raw or cooked; oil from seeds is used in cooking and as medicine	
Manilkara hexandra	Fruit	Ripe fruits are very tasty and liked by all tribal communities	
Olax psittacorum	Ripe Fruit	Ripe fruits are edible	
Phoenix acaulis	Ripe Fruit	Ripe fruit with scanty flesh is tasty and edible	
Phoenix sylvestris	Ripe Fruit	Very commonly eaten as a summer fruit by all tribes	
Phyllanthus emblica	Mature Fruit	Ripe and mature fruits are edible as a medicine	

Scientific Name	Edible part	Mode of consumption and use	Edible Fruit
Protium serratum	Ripe Fruit	Ripe sour fruits are edible and also made into pickles by some tribes	
Schleichera oleosa	Ripe Fruit & Seed	Ripe fruits are edible, seeds are eaten raw, seed oil used as a vegetable	
Semecarpus anacardium	Fleshy thalamus, seed	Hypocarp is eaten when ripe; Kernel also edible; Kernel oil as a wood preservative	
Solanum torvum	Tender Fruits	Young fruits fried with ghee or oil is eaten as a vegetable	
Spondias pinnata	Mature Fruits	Ripe fruit is directly edible; green fruits as a vegetable; are also made into pickles and chutney	
Syzygium cerasoides	Ripe Fruits	Mildly acidic ripe fruits with scanty pulp are eaten; also medicinally important	

Scientific Name	Edible part	Mode of consumption and use	Edible Fruit
Syzygium cumini	Ripe Fruits	Ripe fruits are widely used often after treating in salt water; possess medicinal properties.	
Tamilnadia uliginosa	Ripe & Immature Fruit	Ripe fruits are eaten as such or after being boiled, roasted or made into curries	
Uvaria hamiltonii	Ripe Fruit	Fully ripe fruits are quite tasty and eaten by tribal people	
Xylia xylocarpa	Seed	Seeds are removed by breaking the fruit cover and eaten after roasting	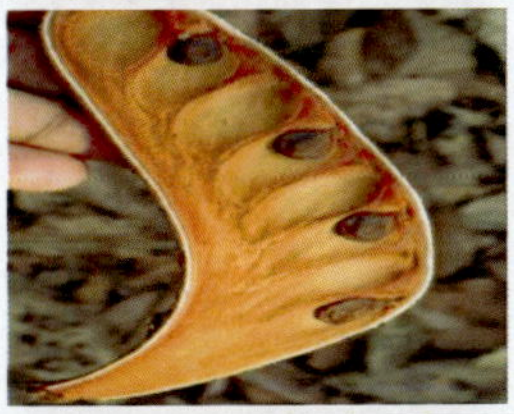
Ziziphus funiculosa	Mature Fruit	Fruits after maturity are eaten	
Ziziphus nummularia	Ripe Fruit	Ripe fruits are eaten largely by children	

Scientific Name	Edible part	Mode of consumption and use	Edible Fruit
Ziziphus oenoplia	Ripe & Mature Fruit	Very common fruit collected and consumed by children; fruit used by Munda tribe for stomach pain.	
Ziziphus rugosa	Ripe & Mature Fruit	Ripe fruits being sweet are eaten.	

Suggested Readings

Azamal Husen Rakesh Kumar Bachheti and Archana Bachheti 2021 *Non-Timber Forest Products: Food, Healthcare and Industrial Applications,* New York: Springer

Parthasarathy VA Utpala Parthasarathi and Kumar A 2008 *Spice Crops Tree Spices Vol. 2* New Delhi: Today and Tomorrow Publishers

Kumar N 1997 *Introduction to Spices, Plantation Crops, Medicinal and Aromatic Plants,* Delhi: Oxford and IBH Publishing

CHAPTER 11
Bamboos, Canes and Grasses

C.N. Hari Prasath and M.P. Divya
Forest College and Research institute, Mettupalayam

BAMBOOS

Bamboos, the most diverse group of plants in the grass family belong to the subfamily Bambusoideae of the family Poaceae (Gramineae). Bamboos are fast growing perennial plants and are found in the tropical, subtropical and temperate regions of the world. The geographical distribution of bamboo largely depends on climatic factors such as precipitation, temperature, altitude and soil conditions. Large tracts of natural bamboo forest are found in tropical Asian countries between 15° and 25° North latitudes. In India bamboo is found naturally almost throughout the country except in the Kashmir region.

As per the FAO report, there are about 1600 species of bamboo in 90 genera across the world. As per the FSI 2021 report, bamboo forests have increased in size from 13,882 million culms (stems) in 2019 to 53,336 million in 2021. India has about 125 indigenous and 11 exotic species of bamboo from 23 genera. Bamboos are found in abundance in the deciduous semi-evergreen forests of the northeastern region of the country and the tropical moist deciduous forests of northern and southern India. The major bamboo genera found in India are *Arundinaria, Bambusa, Chimonobambusa, Dendrocalamus, Dinochola, Gigantochloa,* etc. The North Eastern states and West Bengal account for more than 50% of the bamboo resources in the country. Other bamboo rich areas of the country are the Andaman & Nicobar Islands, Chhattisgarh, Madhya Pradesh and Western Ghats. Although bamboo is found in almost every state of the country, its distribution and concentration vary primarily due to climatic and edaphic conditions. *Bambusa* and *Dendrocalamus* are the species of tropical conditions, whereas *Arundinaria* occurs in the temperate region and are common at high altitude in the Western and Eastern Himalayas. *Phyllostachys* is a genus of temperate to warm temperate regions. *Dendrocalamus strictus* is a predominant species in dry deciduous forests, while *Bambusa bambos* thrives in moist deciduous forests. *Gigantochloa rostrata* is the most important bamboo species in the semi-evergreen of the Andaman Islands. The commercially important bamboo species of India are given in Figure 11.1 (Fig. 11.1).

Bambusa bambos

Dendrocalamus strictus

Bambusa vulgaris (green)

Bambusa tulda

Bambusa balcooa

Dendrocalamus giganteus

Fig. 11.1 Important Bamboo species

Bamboo culms grow from the dense root rhizome system. There are two types of rhizomes *i.e.,* monopodial and sympodial. The monopodial rhizome grows horizontally and the buds develop either upward, generating a culm, or horizontally, with a new track of the rhizome net. Some bamboos are non-clump forming with a new track of the rhizome net. Some bamboos are non- clump forming with culms distant from each other and can be invasive. They are generally found in temperate regions and include the genera *Phyllostachys* and *Pleioblastus*. In other species, the sympodial rhizomes which are short and thick and the culms, which are above the ground are close together in a compact clump, which expands evenly around its circumference. Their natural habitat is tropical regions and they are not invasive. The main genera are *Arundinaria, Bambusa* and *Dendrocalamus.* Bamboo contributes to the subsistence needs of about 2.5 billion people around the world, a majority of whom are tribals, forest dwellers, or communities dependent on forest resources. It is estimated that in India, there are about 2 million traditional artisans whose livelihood depend on harvesting, processing, value addition and selling of bamboo products such as baskets, mats, handicrafts, etc. Major advantage of bamboo is its versatility in making a variety of products by small entrepreneurs without any major initial investment.

Bamboo thrives in a wide range of climatic and edaphic conditions. Owing to its worldwide distribution, bamboo plays a very important role in biodiversity conservation, carbon sequestration and soil moisture conservation. The various properties of bamboos are available in different sizes, lightweight, strong, hard, flexible, straight, fast-growing, abundant and hence have many uses such as in making normal and fine quality paper, fishing poles, furniture, flooring, handicrafts, walking sticks, etc. Young bamboo shoots are used as vegetables in some cuisines. The government of India promulgated the Indian Forest (Amendment) Ordinance, 2017 to exempt bamboo grown in non-forest areas from the definition of a tree, by amending the section 2(7) of the Indian Forest Act 1927 and thereby doing away with the requirement of felling/ transit permit for its transport and economic use. The major objective of the amendment is to promote the cultivation of bamboo in non-forest areas to achieve the twin objectives of increasing the income of farmers and also increasing the green cover of the country.

Important Bamboo Species

i) *Arundinaria racemosa*

It is a shrub, erect and gregarious bamboo with long rhizomes. Culms are 3-4m high, 3-5 cm in diameter and yellow with prominent nodes. Culm sheath is

6 x 25 cm; leaves 10 x 1.0 cm linear-lanceolate. It is found in North Bengal, Sikkim and Arunachal Pradesh at an elevation of 3000-3660 m. The species forms dense undergrowth in temperate forests. It is used for roof construction and matting for houses, fencing and garden support.

ii) *Bambusa balcooa*

It is a tall caespitose bamboo with culms 16-23 m high and 10- 15 cm in diameter. The culms are green, thick-walled, branched from the base, nodes thickened with whitish ring above and hairy below. Branches from the lower nodes are leafless, hard and sometimes thorn-like. Culm sheaths are deciduous, leaves lanceolate 15-30 cm x 2.5 cm in size. It occurs in Bihar, West Bengal, Tripura and Assam. It is cultivated in the foothills of Arunachal Pradesh and has been introduced in Kerala and Uttar Pradesh. The species is reported to flower gregariously. Sporadic flowering also takes place. The species can be propagated by seedling and rhizome. Culm and branch cuttings are also reported to root when treated with 100 ppm IAA. It is probably the best and the strongest bamboo species for building purposes and is mostly used for scaffoldings. In Tripura, it is used for making incense sticks (agarbatti) and also used in making huts and houses in rural areas.

iii) *Bambusa bambos* syn. *B. arundinacea*

The species prefers rich and moist soil and grows along the perennial rivers and moist valleys. It attains the best growth in moist deciduous forests receiving rainfall of nearly 2000 mm. It is a thorny bamboo, bearing bright green shining culms, 20-30 cm high and 10-18 cm in diameter. The nodes are swollen and lower nodes produce aerial roots. Internodes are 20-40 cm long. Branches develop from the nodes from the base upwards, lower branches are having curved spines. Culm sheath 20-35 x 20-30 cm coriaceous with dark brown hairs. Leaves are variable in size usually linear or lanceolate. The species flowers sporadically and also gregariously. The cycle of gregarious flowering is reported to be 30-50 years. After gregarious flowering, sufficient quantity of seeds can be collected. A single clump may yield about 50-100 kg of seeds.

The seeds have short viability. However, seeds stored properly may have viability even up to 6 months. The species can be propagated by seedling, rhizome and culm cutting. Seedlings are raised in the nursery and one year old seedlings are planted in the field. Rhizome planting is common in villages. Young shoots of the species are made into various marketable products such as curry, candy, chutney and pickles. In Manipur, young shoots are used as vegetables. Fermented preparation of young shoots known as Soibum is a delicacy in Manipur. Protein content in seeds is comparable to rice and can be eaten. The species produces strong culms which are used for making various handicrafts and constructional items.

iv) *Bambusa nutans*

It is a medium sized elegant bamboo. Culms 7-14 m high, 5-7 cm in diameter, loosely clumped, much branched above and unbranched below. Culms are straight, green, smooth and internodes 25-45 cm long. Leaves 15-25 cm long, 2 to 3.5 cm broad, linear-lanceolate and acuminate at the apex. Young shoots yellowish green at the apex. This bamboo occurs naturally in the sub- Himalayan tract from the Yamuna to Arunachal Pradesh between 600-1500 m altitude. It is commonly cultivated in Northern Uttar Pradesh, Odisha and North Bengal. The species grows well in the moist hill slopes and flat uplands in well drained loamy to clayey loam soils. Apart from sporadic flowering, this bamboo seems to flower gregariously at 35 to 40 years cycle. The species can be raised as seedlings. In several areas, it is mostly by planting offsets, sometimes rooting of culm cutting is also done. It is an elegant bamboo and is largely grown for ornamental purposes. The culm is hard, strong, straight and mainly used by local people for poles and other purposes.

v) *Bambusa pallida*

It is an elegant bamboo growing in thick clumps. The culms are 13-20 m high, 5-8 cm in diameter, smooth, olive green, nodes not prominent and internodes 40-70 cm long. Leaves 10-20 cm long, 1-2 cm broad, linear-lanceolate, rounded at the base with scattered long bristles, petiole very short. Young shoots are spear-shaped. The species occur in northeast India in Arunachal Pradesh, Assam, Northern Bengal, Meghalaya, Mizoram, Sikkim and Tripura. It occurs in hills mainly between 700-2000 m altitude. It is also cultivated in plains. The species is generally used in building construction and making baskets and mats. It is extensively used for the manufacture of toys, wall plates, screens, wall hangers, etc.

vi) *Bambusa polymorpha*

It is a large evergreen densely tufted, erect or drooping bamboo with tall clean culms, attaining a height of 24 to 27 m and diameter of 15-20 cm. Internodes are 40 to 60 cm long. Culm sheath 20-25 cm long, 20-30 cm broad and usually persistent. Leaves 7-18 cm long, 1-2 cm broad linear-lanceolate, base rounded, petiole 2-3 mm long. Young shoots are brownish-green. This handsome bamboo is a native of Arunachal Pradesh, Manipur, Meghalaya and Tripura. It is also found in Madhya Pradesh, Kerala and is cultivated in North Bengal, Karnataka and Tamil Nadu. It grows well in low hill slopes along the valleys in deep fertile well drained riverine alluvium. It has been reported to flower gregariously and sporadically. Gregarious flowering takes place usually in a cycle of 40 to 60 years. It is used for making paper and pulp, *agarbati* sticks, fibre boards, toys, mats, etc. It is considered to be the best for walls, flooring and roofing. Young shoots are used for edible purposes. Young shoots are sweet and eaten raw as well as widely used for curry, pickles, chutney, etc. It is a graceful species for landscaping.

vii) *Bambusa striata* syn. *B. vulgaris*

It is a moderate sized bamboo, not densely tufted. Culms are 10-20 m high, 5-10 cm in diameter, bright green, erect and yellowish with thick walls. Nodes are prominent, often with a narrow ring of internodes up to 45 cm long. Culm sheaths are 15-25 x 20-35 cm rounded and truncate at the top. Leaves narrowly or broadly lanceolate, 15-25 x 2-4 cm in length. It is cultivated extensively in the northern part of the country particularly in northern east Madhya Pradesh, Bihar, Uttar Pradesh and other states of the country mainly for ornamental purposes. It flowers both gregariously and sporadically. It is raised by seedling and rhizome planting. It is used for making paper, scaffolding poles and other structures. Culms are durable and fairly compare with timber. Rings prepared from the split culms are put into ear perforations by the Tunkul-Naga tribe of Manipur. *B. vulgaris* var. *striata* is an elegant tufted bamboo, culms 4-8 m high,5-8 cm in diameter, glabrous, yellow with light green stripes, shining, internodes 10-15 cm long. It is generally cultivated for ornamental purpose. Culms are used as poles for construction purpose. It can be used for pulp and paper.

viii) *Bambusa tulda*

It is an evergreen to deciduous tufted and gregarious bamboo with culms usually 15-23 m high and 7-10 cm in diameter. Culms are glabrous green when young and grey green on maturity. Nodes are slightly thickened and internodes are 40-70 cm long. Culm sheaths are 15-25 cm long and 10-20 cm broad and usually deciduous. Leaves 15-25 cm long, 2-4 cm broad, lanceolate, apex acuminate and yellow stripes on the green surface. The species is naturally distributed in Bihar, Assam, Meghalaya, Mizoram, Nagaland and Tripura. It is cultivated in Arunachal Pradesh, Uttar Pradesh, North Bengal and Karnataka. It flowers gregariously and sporadically. The life span of seeds of this species is not much, the viability decreases after one month period. Freshly collected seeds show better germination, germination starts within 7 days and completes within 20 days. It is used for house construction and scaffolding. Tender shoots of 0.6 m high are frequently used for making pickles. It is used by pulp and paper industries. It is used for making toys, mats, screens, wall plates, wall hangings, hats, baskets, food grain containers, etc. In Arunachal Pradesh, the flute used by the priests is made from this species with a belief that it keeps the evil spirit away. Fishing rods are also made from this species.

ix) *Dendrocalamus giganteus*

This is the tallest bamboo with close culms and slender branches. Culms are 25-30 m tall, 20-30 cm in diameter, usually 2-5 cm thick-walled and dull green. Culm sheath is 30-50 cm long and 15-20 cm broad. Leaves are variable in size, in the main culm up to 50 cm long and 10 cm broad. It is a native of Burma and introduced to India in 1930. Presently, it is cultivated in Arunachal Pradesh, Assam, Manipur, Nagaland, West Bengal and occasionally in other parts of the country. It is good for moist hill slopes and flat lands with rich loamy soils. It flowers both gregariously and sporadically. Gregarious flowering in some clumps has been observed in certain parts of the country. It is propagated by seeds and rhizomes. Rooting of culm cutting is also reported to be successful. This bamboo is primarily used for hut and house construction, the mast of boats, flower vases, buckets and other decorative items. In Arunachal Pradesh, it is used as a water pitcher, particularly by Abor and Mishmi tribes. Young shoots are considered to be a good vegetable. It is a good material for paper and pulp. It is also used for making various handicraft items.

x) *Dendrocalamus hamiltonii*

It is a large caespitose bamboo often culms at an angle or is curved downwards. Culms 15-20 m tall, naked below, much-branched above. Culm sheaths are long and stiff variable in size. Leaves are variable in size, with larger leaves (35.0 x 3.7 cm) on new shoots. It is distributed in northwest Himalayas, Sikkim, Bhutan, Arunachal Pradesh, Assam, Manipur, Meghalaya, Mizoram, Nagaland and Tripura. It is generally used for walling of native huts, other constructions, basket making, mats, water and milk vessels, fuel, floats, etc.

xii) *Dendrocalamus hookeri*

It is a large tufted bamboo with culms 15-20 m high and 10-15 cm in diameter. The culms are dark green, internodes 40-50 cm long, rough, hairy with long curving branches. Culm sheath is 40 cm broad at the base and 20-30 cm long. Leaves are large, oblong-lanceolate with a long acuminate tip. Young shoots are covered with black tomentum. It is distributed throughout Arunachal Pradesh, Nagaland, Mizoram, Meghalaya, North Bengal, Sikkim and Manipur in the hills between

600-1500 m altitude. The seeds are obtained in gregarious flowering, clumps die soon after flowering. It can be propagated by seed, rhizome and culm cutting. The species is generally used for building purposes. It is also used for making baskets and water buckets. The species is quite similar to *Dendrocalamus hamiltonii* in qualities.

xiii) *Dendrocalamus longispathus*

It is a large handsome tufted bamboo with culms usually 15-18 m high. The culms are glaucous green when young and turn greyish green when old, nodes are slightly swollen, often rooting and internodes 25-60 cm long. Culm sheaths 35-50 cm long and 10-20 cm broad. Leaves 10-30 cm long, 2.5-4.5 cm broad, oblong-lanceolate. Young shoots are spear-shaped and sheath green. The species is found mainly in Manipur and Tripura. It has also been introduced in West Bengal, Bihar, Assam and Odisha. The species grows in moist localities and is found mainly along streams. The species often flowers sporadically and gregariously after long intervals. Seeds have short viability. Freshly collected seeds have the highest viability. The species can be propagated by nursery raised seedlings and rhizomes. Culms are not very strong and therefore, are considered inferior to other species for building purposes. In Tripura, it is largely used for making baskets and food grain containers. It is a handsome species and is generally planted for ornamental purposes.

xiv) *Dendrocalamus membranaceus*

It is a moderate-sized strong bamboo that forms loose clumps. The culms are straight, 20-25 m high and 8-10 cm in diameter. It is a native of Burma, Thailand and China and has been introduced successfully in Kerala and West Bengal. It occurs in moist forests but it is capable to tolerate dry conditions. Since it is a strong bamboo, it is highly preferred for construction purpose. Its shoots are eaten. It is also used for making chopsticks, shreds and paper.

xv) *Dendrocalamus strictus*

It is a deciduous and densely tufted bamboo with culms 8-15 m high and 4-8 cm in diameter. They are blue green when young and dull green or yellowish when old. The nodes are swollen and the basal node often rooting. Internodes 30-50 cm and thick-walled. Culm sheaths are variable in size, lower ones are usually shorter than the upper. Leaves are shorter in dry localities and larger in moist localities. It is widely distributed in semidry and dry areas up to an altitude of 1000 m. In India, it is a predominant species in Madhya Pradesh, Uttar Pradesh, Maharashtra and Odisha usually associated with tropical dry deciduous forests. It flowers both sporadically and also gregariously (40-60 years cycles). The clumps die soon after gregarious flowering. The seeds can be stored for about one year. Their viability can be retained for a longer period if kept under optimum temperature. The species is propagated by seed, seedlings and rhizomes. Culm cutting has also been reported successful under controlled conditions. This is one of the most important bamboo species and is widely used for construction purposes in rural housing. It is the most common bamboo used for pulp and paper. Its young shoots are pickled and made into delicious curry, preferred by tribals. The thin solid culms are used as lathis (walking sticks) in rural areas. It is also used for making baskets and handicraft items. Leaves are good fodder. The decoction of leaves is used as medicine; silicious matter is used as a tonic and astringent. The species is largely used in the afforestation of wastelands.

xvi) *Drepanostachyum falcatum* syn. *Arundinaria falcata*

It is a low level bamboo of the western Himalayas distributed from Ravi to Nepal at an elevation of 1200-2230 m. This species grows gregariously forming an undergrowth in the forests of deodar, oak, etc. in moist shady locations. It forms a large number of culms, sometimes more than a hundred culms in a clump. The culms are 4m high, 1-2 cm in diameter, smooth and cylindrical. Culm branches are numerous. Culm sheaths are 17-30 x 3-7 cm size and papery straw in red colour. Leaves are 8 x 0.5 cm in size, linear-lanceolate and much larger in young shoots. Flowering is both sporadic and gregarious. It is used for basket work, fishing rods, arrows, hookah pipes, etc. It is especially prized for fishing rods. Leaves are used for thatching, roofing, etc. Young shoots are edible.

xvii) *Gigantochloa albociliata* syn. *Oxytenanthera albociliata*

It is a densely tufted bamboo almost evergreen with culms 6-9 m high and 1.5-2.5 cm in diameter. The culms are greyish green with white strips. The young shoots are purplish green. The species is distributed in Burma and widely cultivated in West Bengal, Assam, Arunachal Pradesh, Meghalaya and elsewhere in the country. It flowers sporadically. However, the seeds from the sporadic flowering are reported to be infertile. It is locally used by villagers for household purposes.

xviii) *Gigantochloa rostrata* Syn. *Oxytenanthera nigrociliata*

It is a tufted dark green bamboo with culms 5-8 m tall and 2.5-5 cm in diameter. Culms are thick-walled, the basal portion with yellowish stripes. The species is distributed in Assam, Meghalaya, Odisha, Bihar, Madhya Pradesh, Maharashtra and Karnataka. It prefers growing in moist localities along streams. It is used for building huts, making baskets and also as a raw material for paper and pulp industries.

xix) *Indocalamus wightianus* syn. *Arundanaria wightiana*

It is a shrubby and erect bamboo. Culms 3 m in height with a bulbous base and golden hairs. Internodes are flat on one side, nodes swollen with a hairy ring below. Culm sheath is papery and straw colored. Leaves are 13.5 x 2 cm in size; ovate-lanceolate. It occurs in Nilgris, Palghat and Tirunelveli in South India between an elevation of 1800-2600m. It forms gregarious undergrowth in the evergreen sholas. Culms are 2 to 3m high. It flowers annually.

xx) *Melocanna baccifera* syn. *M.bambusoides*

It is an evergreen bamboo, the culms are 10-20 m high and 4 to 7 cm in diameter, green when young and straw colored when old. Culm sheath 10-15 cm long yellowish green when young and yellowish brown when old. Leaves are 15-30 cm long, 2.5-5 cm broad oblong-lanceolate with an acuminate apex. The species is distributed in Assam, Manipur, Meghalaya, Mizoram, Tripura and West Bengal in lower hill slopes and plains. In Mizoram and Manipur, this species occupies a large area of abandoned shifting cultivation. It is propagated mainly by rhizomes as they are vigorous and highly productive. This is due to a large rhizome neck. Sporadic and gregarious flowerings have been observed in this species. The species produces the largest fruit among bamboos. The size of fruits is generally 10 cm x 7.5cm excluding the beak, which is about 2.5cm long. The species is employed in various uses. It is the principal material used for building houses for making woven wares and an important source of superior paper and pulp. The species is also used for making toys, mats, wall plates, screens, wall hangings, hats, umbrella sticks, baskets, food grain containers, etc.

xxi) *Ochlandra scriptoria* syn. *O. rheedii*

It is a shrubby bamboo with culms up to 5m in height and 2.5cm in diameter, found gregarious at low elevations in Kerala. It is also cultivated along with the skirts of ponds and paddy fields as a soil binder. This species is mostly used for making mat baskets and pulp and paper.

xxii) *Ochlandra travancorica*

It is a shrubby bamboo found in Mysore, Travancore, Anamalai and Tirunelveli hills up to 1550 m elevation. Culms are 6m in length and 5cm in diameter with rough, thin-walled, solid internodes up to 1.5m in length. It is used for making agricultural implements and tool handles. It is also used for making temporary huts and thatching.

xxiii) *Pseudostachyum polymorphum*

It is thin-walled, shrubby bamboo found along river banks and valleys in Sikkim, extending eastwards to Assam. It is in considerable demand among tea planters, as it gives the best materials for making baskets.

xxiv) *Pseudoxytenanthera ritcheyi*

It is an erect, thick walled bamboo occurring from the Konkan to Anamalai hills on the Western Ghats. It is often cultivated. The culms are 6m in height and 2.5 cm in diameter. They are strong and used for tent poles, walking sticks, baskets and umbrella handles.

xxv) *Schizostachyum pergracile*

It is widely distributed in Bihar, Assam, Naga Hills, Madhya Pradesh, Odisha and Andhra Pradesh. It is an elegant deciduous tufted bamboo with thin walled glaucous green culms, attaining a height of 9 to 12 m or more and a diameter of 5-7.5 cm under favorable conditions. It is cultivated in fresh loam soils and is remarkably gregarious. This species often flowers sporadically and occasionally flowers gregariously over extensive areas. It is used in mats, fishing rods, hats and other handicraft items.

xxvi) *Thamnocalamus spathiflora*

It is densely tufted shrubby bamboo. Culms 6 m in height, 2 cm in diameter, smooth, glaucous green when young and turns shining yellow with age. Nodes much raised dark colored, internodes 20-25 cm. Culm sheath 13 x 15cm, loose, light yellow. It is a small sized bamboo, growing gregariously in the undergrowth of fir, spruce and deodar of western Himalayas at 2100-2700 m elevation. The culms are used in making pipes, mats, baskets, etc., It occurs in the western Himalayas from Sutlej throughout Nepal to Arunachal Pradesh between 2100-3050 m elevation above mean sea level.

GRASSES

Grasses occupy wide tracts of land and they are evenly distributed in all parts of the world. They occur in all types of soil in all kinds of situations and under all climatic conditions. In certain places, grasses form a leading feature of flora. As grasses do not like shade, they are not usually abundant within the forests either as regards the number of individuals or species. But in open places, they do very well and sometimes whole tracts become grasslands. On account of their almost universal distribution and their great economic value grasses are of great importance to human. The grasses growing in the pasture land and the cereals grown all over the world are of more value to human and his domestic animals than all the other plants taken together (Fig. 11.2).

Saccharum spontaneum

Cenchrus ciliaris

Dichanthium annulatum

Panicum antidotale

Hyparrhenia tamba

Themeda arundinacea

Fig. 11.2 Important grass species

Suitable Grasses for Paper Making

i) Bhabar / Sabai grass (*Eulaliopsis binata*)

Sabai grass is a perennial grass that is found in the bare slopes and forest blanks of the Sub-Himalayan tracts and Bihar, Odisha, West Bengal, Madhya Pradesh and the eastern parts of Punjab. It is chiefly used in paper making. Annually 60,000 to 80,000 tonnes of this grass are supplied to paper mills; the principal sources of supply being located in Uttar Pradesh, Bihar, Odisha and Madhya Pradesh. The grass is collected and usually pressed *in situ* and sent to the paper mills. The chief areas of consumption lie in Uttar Pradesh and the Punjab where adequate supplies of bamboo are not available. At one time the supply was insufficient to meet the demand of the paper mills and cultivation of this grass in a suitable place was contemplated. However, the position has eased with the increasing use of bamboos for paper-making. It has been found that *Imperata cylindrica* and *Desmostachya bipinnata* can be used in admixture with sabai grass to produce a satisfactory paper.

Other suitable grasses

The following grasses yield the first-class material for paper pulp: *Saccharum spontaneum, S. munja, Narenga porphyrocoma, Arundo donax, Phragmites karka, Themeda arundinacea* and *T. nillosa.*

Fodder Grasses

i) *Andropogon* spp.

Andropogon genus is an important and widespread genus of the perennial genus, particularly in dry regions of India and is a principal constituent of wild forage. *Andropogon pumilus* is a gregarious annual grass, attaining a height of 0.6 m and is found throughout the drier parts of India, especially in the arid region of Rajasthan and Gujarat. It is considered to be a good fodder and is relished by cattle either green or when made into hay. Five separate cuttings can be made annually between September and January, producing about six tonnes per hectare. *A. lividus*, locally known as purple grass from the colour of the inflorescences, is a valuable fodder grass found in the hills of Tamil Nadu. *A. ascinodis* is gregarious in the hills of Assam. *A. gayanus* is a perennial species, found in the swampy regions, commonly called gamba grass. It is a valuable hay and green fodder grass in most part of African regions. It is palatable and eaten by cattle when young but up to flowering stages. Once a flowering panicle appears, it becomes hard in structure and of low nutritional value.

ii) *Apluda mutica*

It is commonly known as Mauritian grass and is a rambling perennial grass with stem upto about 3 m long, rooting from the lower nodes. This is considered to be a fairly good fodder grass and is readily eaten by cattle when young. It is very common in the plains and at low elevation in the Himalayas. In hedges and bushy places, it usually assumes a climbing habit. It often constitutes a large part of the undergrowth in forests. It is considered to be a good fodder in young stage.

iii) *Bothriochloa* spp.

Bothriochloa is a common and widespread genus of plants in the grass family and 12 species of *Bothriochloa* are found in the Indian region. *B. ischaemum* is a warm seasoned perennial grass, native to Europe, Asia and Africa. The average annual yield of green matter from this grass is fairly high. Young plants which are not too old are readily eaten by cattle. The annual yield of dry grass is about 8 tonnes per hectare. *B. pertusa* is a perennial grass attaining a height of 30 cm, commonly growing in black cotton soil and is considered a good fodder grass, both for grazing and stacking and suitable for hay and silage.

iv) *Bromus* spp.

Bromus is a large genus of grasses, classified in its own tribe Bromeae. They are commonly known as bromes, brome grasses, cheat grasses or chess grasses and are found in mountains of tropical and temperate regions. More than 10 species of the genus are found in India in the Himalayas and other temperate mountain regions. *B. asper, B. inermis,* and *B. catharticus* are good and palatable pasture grasses found in the high mountain areas of India. The average yield of *B.inermis* is about 8.25 tonnes per hectare. It is cultivated as a forage plant in drier areas and also is a drought tolerant species. It is one of the most palatable pasture grasses in temperate region.

v) *Cenchrus* spp.

Cenchrus is a widespread genus of plants in the grass family. It is native to many countries in Asia, Africa, Australia, the America and various oceanic islands. It comprises eight species growing in India. The important and common Cenchrus species are discussed below:

a) Cenchrus biflorus

Cenchrus biflorus is an annual grass of the family Poaceae with culms between 4-90 cm high and is commonly found in the arid regions of India. It appears in bushes and as a weed on degraded land. *Cenchrus biflorus* is also a valuable fodder plant for ruminants, particularly at its early stages of development. It is readily eaten by sheep in the pre-flowering stage and after the seed bur has been shed.

b) Cenchrus ciliaris

It is commonly known as African foxtail grass and is native to tropical Africa, the Mediterranean region and the hotter and drier parts of Asia. It is a deep rooted grass, tolerates drought and will grow at an altitude up to 2000 m. It is a perennial tufted grass, 18 to 45 cm high, erect or decumbent. In India, it is a natural species largely found in Rajasthan, Haryana, Punjab, Gujarat and parts of western U.P and Tamil Nadu. It is resistant to drought and stands cutting well, giving high yield in areas having good rainfall and well drained soil. For raising a fresh pasture, seeds are sown at the rate of 18 to 22.5 kg per hectare. The average annual yield of this grass is 20 to 30 tonnes of green fodder per hectare. This grass is highly nutritious compared to other fodder grasses and it can be fed green or made into hay or silage.

vi) *Chrysopogon* spp.

Chrysopogon is a genus of tropical and subtropical plants in the grass family. They are widespread across Eurasia, Africa, Australia, south-eastern North America and various islands. It yields high quality valuable fodder and is found almost throughout India. There are 14 species of this genus that occur in India, of which some species are valued as fodder. *Chryosopogon aciculatus* is a common grass in pasture and in lower hills. *C. fulvulus* is a perennial, tufted grass growing in open dry and stony regions of West Bengal, Meghalaya, probably north and north-west India. It is highly valuable and nutritious before flowering and it can be converted into hay.

vii) *Cynodon dactylon*

It is a hardy, perennial grass, forming thick, spreading mats on and under the surface of soils, greatly varying according to soil and climate. It is found throughout India up to 2000 m elevation and is considered as weed in agricultural areas. It is a valuable pasture grass, which remains green during hot weather. It is fed to cattle green and is also suitable for hay. Normally, four cuttings are made in a year; each cutting yields about 8 tonnes per hectare. It serves as pasture forage, particularly during the summer and hay is a good source of vitamin A for beef cattle.

viii) *Dichanthium annulatum*

Marvel grass (*Dichanthium annulatum*) is a tropical grass originally from North Africa and India that is used for pasture in tropical and subtropical zones. It is particularly used in India. It is a densely tufted perennial, 60-100 cm high and common throughout the plains and hills of India up to 500 m elevation. It is very common among wild fodder grasses. The grass is eaten by cattle both when it is young and in flower. It makes good hay and is suitable for silage. It is an excellent fodder grass. It stands cutting well and yields up to 9 tonnes per hectare. It makes good hay if cut before flowering.

ix) *Eragrostis* spp.

A wide genus of annual or perennial grasses, of which 25 species are found in India. Some of the species are considered to be good fodder grasses viz., *E.abyssinica, E.ciliaris, E.curvula, E.plumosa, E tenella,* and *E.tremula*. *E. trenella* is found throughout India up to a height of 1200 m and is eaten by cattle, either green or as hay. *E.tremula* and *E. cilianensis* are other popular species. A few exotic grasses have been successfully introduced for soil conservation purposes. Of these, *E.curvula* has proved to be very useful and is a very important fodder grass.

x) *Panicum antidotale*

Blue panic grass is a vigorous, tufted perennial grass, 1.5 m to 3 m high throughout the Himalayas and the Upper Gangetic Plain and specifically in various regions of the Indian state of Punjab, Tamil Nadu, and north-west India. It is grazed by cattle as it is palatable only when young. *Panicum antidotale* can yield 10-50 tonnes of grass/ha. Hay yields vary from 2.5 to 6 t DM/ha under rainfed conditions to 5 t/ha under irrigation. It provides a palatable pasture and is valuable as a fodder crop for light sandy soils in dry areas. It is readily grazed by livestock before flowering. At flowering, the spikes have hard and woody stalks that decrease palatability rapidly.

xi) *Heteropogon contortus*

It is a tropical, densely tufted gregarious perennial tussock grass and found mostly in the dry places throughout India and also in Himalayas up to 1500 m elevation. It is mainly used as fodder when it is young. It can be grazed or cut for hay or silage. The sharp callus of the fruits aided by the awns is apt to cause severe abscess in animals but when young, the grass makes very good fodder. An average yield is about 50 tonnes per hectare. The silage prepared from young grass is readily consumed by animals

Thatching Grass

Thatching grass has been used as a roofing material from the earliest times. Thatching grasses are used throughout India for roofing "Katcha" buildings in rural areas. The grass for this purpose is usually cut as soon as it is mature. If left uncut after it has matured, it deteriorates and rots. It should be cut regularly every year as the clumps fall off in quality and quantity if uncut for a season. If not cut it may be burnt over, which has the same restorative effect as cutting. Grass culms are to be at least 1 m long, preferably longer than 1.5 m is preferable for thatching purposes.

The commonest grass for thatching is *Imperata cylindrica* which covers enormous areas all over India and also thatching the roofs of traditional homes throughout south east Asia. Another species valued for this purpose is *Saccharum spontaneum*, which has superior durability. The other thatching grasses are *Heteropogon contortus, Saccharum bengalense* and *Erianthus ravennae.*

Preparation of Thatching Grasses

a) Cleaning and bundling

After cutting and loosely bundling, each bundle is shaken vigorously to dislodge all loose material. The bundles are then cleaned by passing a sickle through them, working towards the butt. This removes the remaining leaf growth from the lower two-thirds of the stalks. The grass is then remade into bundles about 1 to 1.5 m long and between 75 and 100 mm in diameter. These bundles are tied with a thong of twisted grass or with twine and packed in heaps (round pyramids) about 2 m high and 2.5 to 3 m in diameter at the base.

b) Combing

A comb is made by driving several 75 x 3.5 mm diameter round wire nails into a 300 mm length of the horizontal pole. The nails are spaced at about 12 mm in a straight line. The bundles of grass are placed across the top of the comb and pressed down so that the stalks are separated by the nails. The bundle is then pulled through the comb from the top to the butt end. After combing, the bundles should be stacked clear of the ground and undercover. They are normally baled for transportation in batches of 10 – 20 depending on the size of the bundles.

Grasses for ropes

Eulaliopsis binata and *Desmostachya bipinnata* can be made into rough ropes for use in country cots. In places, these fibres are made into large fishing nets. The fiber obtained from the leaf-sheath of *Saccharum bengalense* is used for manufacturing cordage and ropes. Other grasses yielding useful cordage fibre are *Themeda arundinacea* and *Saccharum spontaneum*.

CANES

Canes (Rattans of commerce) are the stems of climbing palms of the genera *Calamus* principally and of a few other related genera. The genus *Calamus* consists of about 390 species and of which, 30 species occurring in India, distributed chiefly in the Himalayas, Assam, Kerala, Karnataka, Tamil Nadu and Andaman. The genus belongs to the family Palmae and is generally found distributed in the virgin forests of tropical and subtropical regions. Most of the species are climbers and twining shrubs that climb over forest trees with the help of hooked spines or by the flagellum-like prolongations of the leaf rachis. These spines are located either on leaves or leaf sheaths. The stem of many species of *Calamus* from the common canes is commercially called rattans. Few other related genera like *Daemonorops, Ceratolobus, Plectocomia,* and *Korthalsia* also yield canes but are of minor importance. Many *Calamus* species are climbing species, their stems are long usually cylindrical having a uniform thickness. They are solid, yellow coloured by spiny leaf sheaths, flexible, elastic and strong. The outer surface is hard and shiny. The shining appearance of the outer surface of the stem is due to the deposition of silica on the surface. The core of the stem is spongy. Nodes are less dominant internode vary in length from species to species and sometimes even in different plants of the same species. Leaves alternate, leaflets are acuminate usually parallel in Fig 11.3 and the rachis modified to the armed flagellum. The culm sheath is armed into spines. Flowers are small and solitary. Fruits are clips and ellipt. Seeds are sub-globose and oblong. The photograph of the important cane species are given in Fig. 11.3.

Calamus rotang

Calamus thwaitesii

Calamus pseudotenuis

Myrialepis paradoxa

Calamus gibbsianus

Calamus vattayila

Fig. 11.3 Important cane species

Important Cane Species

i) *Calamus acanthospathus*: This is a common climber occurring in eastern Nepal, Sikkim, Bhutan and Assam ascending to 2000 m altitude. Stem slender, scandent, as thick as a swan's quill. The internodes are 15 cm long with a diameter of 2.0-2.5 cm, leaves 60-90 cm, leaflets 25-40 cm by 5.0-7.5cm margin naked or spinulose, both surfaces naked or very rarely with a few small lines on the costae of the upper surface with erect, armed short recurved spines. It is used for making handicraft items. It yields strong canes.

ii) *Calamus andamanicus*: It is found in Andaman and Nicobar Islands. The stem is lofty, scandent and as thick as an arm (with the sheaths on). Leaflets are 60-75 cm x 25 cm broad. It is used in making furniture, handicraft items, wickerwork, basket, etc.

iii) *Calamus erectus*: It is distributed in Sikkim, Himalayas, Assam, Khasi Hills and Manipur, ascending to 1200m altitude. Stem densely tufted, 4-6 m high, internodes 5.5-7.5 cm long, 2.5-3.0 cm diameter. Leaves 1.50m length, leaflets few, 40-60 x 3.0-5.0 cm, 1-nerved. Spadices 30-60 cm, spikes 15-25 cm x 1.0 cm across the imbricating coriaceous spathes. Calyx acutely 3-toothed; corolla thrice as long. Stamen-free, filaments broad and erect. Fruit about 2.5 cm long, ovoid-oblong, subtended by the spreading perianth, scales trapezoid, dark brown with a yellow base. The seeds are used as a substitute for betel nut (Areca nut). The young and tender shoots are also eaten, boiled and made into curries and pickled.

iv) *Calamus guru*: This is a slender cane 0.6-0.8 cm in diameter, distributed in West Bengal and Assam.

v) *Calamus flagellum*: It is distributed in Sikkim, Himalayas, Assam and Khasi hills, ascending to 1200 m elevation. Stem 2.5-4.0 cm in diameter. Leaves 2.0-2.5 m; leaflets 60-75 cm x 2.5- 4.0 cm broad, petiole 30-45 cm. spathes tubular, lowest 45-60 cm compressed, upper lacerate. Spadix 2.5-3.0 m, spikes 20-25 cm, pendulous, unarmed, spathes like *Calamus erectus*. Flowers 1 cm long, decurved. The species is used for making furniture and other handicraft items.

vi) *Calamus floribundus*: It is found in Assam, Khasi (Meghalaya) and Mishmi hills. The stem is as thick as the middle finger. Leaflets bright green, 20-50 x 2.5-4.0 cm; spines on rachis 1.0-2.5 cm long, rather flattened and deflexed. Lower spathe 45 cm, coriaceous, terete, mouth very oblique. It is used for furniture and other handicraft items.

vii) *Calamus gracilis*: It is distributed in the Khasi hills, ascending up to 1200m altitude; upper Assam and Cachar. Stem slender, scandent, naked stem not thick.

Leaves 60-90 cm, leaflets 12-20 cm, margins obscurely bristly except at the tip where the bristles are long, petiole very short and rachis fugaciously scurfy. It is used widely in making furniture and handicraft items.

viii) *Calamus latifolius*: It is distributed throughout the Sikkim Himalaya and Assam ascending up to 600m altitude in the Himalayas. Stem stout, 30-60 cm, as thick as a thumb. Leaves with the flagellum 3-5m, leaflets 30-60 cm by 2.5-5.0 cm, upper sometimes flabellate clustered or connate at the base, cross nerves many and close, petiole short, sometimes bearing the leaflets only far up the flagellum; base 2.0 cm broad, smooth, biconvex, spines are numerous and always scattered. It is used for making furniture, handicraft items, umbrella handles, etc.

ix) *Calamus leptospadix*: It is distributed in Sikkim Himalaya, Khasi hills and Naga hills. The stem is slender and scandent. Leaves 1.0-1.5 cm, leaflets 20-25 x 0.70-1.0 cm, thin, pale green, rachis armed with simple recurved spines. Spathes slender, tubular. Spadices several feet long, branched, below and armed with hooked short spines, branches very long and slender with 20-40 recurved spikelets 1.0-2.5 cm long and 0.3 cm in diameter.

xi) *Calamus tenuis*: This is the common cane of North India, found in the sub-Himalayan tract, tropical Himalaya from Kumaon eastwards, Bengal, Assam, Mizoram, Manipur and Tripura. It is also distributed in Kerala. Stems are very long, scandent, internodes not thicker than a goose-quill. Leaves 40-60 cm, leaflets 20-30 x 0.78-1.0 cm, margins minutely setulose or spinulose, bristles, petiole pale, stout spines with broad laterally compressed conical base. It is used for making mats, baskets, screens, furniture, chair seats, etc. It is of medium thickness, being 0.6-1.12 cm in diameter.

xii) *Calamus thwaitesii*: It is distributed throughout central India along streams. The rachis of the petiole bears stout recurved claws with black tips, the upper spathes bearing scattered spinescent tubercles and in smaller fruits with few pale scales erose or the margin channeled down the center and with a narrow brown marginal band.

xiii) *Calamus travancoricus:* It is distributed in the Deccan peninsula from Malabar to Travancore. The stem is slender and scandent. Leaves 45-60 cm, leaflets 10-15 x 1.0- 2.0 cm, broadest about or above the middle and tapering to a capillary point, shining above, rachis scurfy when young, petiole 10-15 cm, dorsally rounded, margins acute much compressed towards the base and are chiefly spiny. The species is used for making furniture, handicraft items, umbrella handles, etc.

xiv) *Calamus pseudo-tenuis*: It is distributed in the Deccan peninsula; on the Western Ghats from Canara southwards. Stem slender, leaflets 25-45 x 1.0-2.0 cm, upper free; rachis rather stout, petiole short, sheath 2.0 cm in diameter. The stem is used for various purposes viz., furniture items, handicrafts, walking sticks, polo sticks and umbrella handles.

xv) *Calamus viminalis:* This is a thin strong cane, distributed in lower West Bengal, Odisha, Andhra Pradesh, Karnataka and Andaman. It is stout scrambling and climbing species. Leaves 60-90 cm, leaflets 10-25 x2-3 cm, pale green, young and rachis and sheath white-floccose, petiole short and rachis very stout, spines slender 2.0-3.0 cm long, pale hardy flattened. Spathes coriaceous, mouth obliquely truncates, flagella with hooked 2-3 fid claws. Spadix and its spreading branches rather stout internodes (and truncate spathes) 1.0-2.0 cm. It is used for making furniture, frames, polo-wicket and umbrella handles, rough cordage, fibre mat and a variety of other articles.

xvi) *Calamus rotang:* This is found in central and southern India, Deccan peninsula and Ceylon and has slender but strong stem. Leaves 60- 90 cm, leaflets 20-30 x 1.0- 2.0 cm, unarmed on both surfaces or armed beneath only, lateral unarmed on both surfaces. Spadix and its slender tubular spathes unarmed or nearly flattened spines. It is used for making baskets, mats, etc. The fleshly mucilaginous, sweet bitter pulp of the fruits is edible. Decoction of the root is used as a remedy for dysentery.

xvii) *Calamus palustris:* This plant grows in evergreen forests, although it can also thrive in dry evergreen forests, allowing it to be found in practically every section of Thailand. It can grow in the valley as well as up the hillside to the top of the hill. It has a wide range of habitats and can be found near villages in regenerated forests, young secondary forests and old rubber plantations.

Propagation of Canes

a) Propagation by seed

Seed is the best source for propagation, but its availability on a large scale is dependent on several factors. *Calamus* spp. is dioecious and flowers annually. Extraction of canes before flowering and the destruction of the natural habitat of canes also directly affect the seeds. Seeds generally mature in the rainy season. The best period for the collection of ripe fruits is April to June. Fruits should be collected from a mature plant of a height of over 12 m. It is advisable to collect fruits directly from the plants when their colour is changing from green to yellow. Only ripe fruits should be collected since they give good germination. The fleshy mucilaginous sweet-bitter pulp of the fruit (eg. *Calamus rotang*) is edible. This pulp attracts ants and birds (eg. *Calamus tenuis*). Fruits are produced in abundance in inflorescence, generally ripen simultaneously. The size and shape of fruit vary depending on the species. Fruits are covered with vertical rows of reflexed overlapping scales.

The ripe fruits are thoroughly washed and the mucilaginous protective coat is removed. Removal of outer scaly cover and inner fleshy layer from seed ensure good germination. The seeds are soaked in water for about 48 hours to induce fermentation of the fleshy layer. The fleshy part of the seed is removed by rubbing with the hands and then the clean seeds settled at the bottom of the vessel are collected. The seeds can be stored for a week but proper care should be taken to keep the seeds moist as dry seeds fail to germinate. Otherwise, gunny bags containing fruits should be immersed in running water for 24 h till the fruit becomes soft enough for removing the mucilaginous covering of the seed coat, then they have to be gently washed by hand till the seeds are completely separated from the pulp. Generally, cane seeds are known to be viable only for short periods *i.e.,* about 1 to 2 months. The moisture content of the seeds ranging between 40-60 percent appears to be very critical for seed viability as above 60 percent the seeds tend to germinate and below 40 percent become non-viable. It may be difficult to provide suitable storage conditions for maintaining the proper moisture content.

Germination starts 30 to 40 days after sowing. Seedlings grow very slowly and it takes 2 years to reach the plantable size of about 70 cm. Some species e.g. *Calamus hookerians, Calamus pseudotenuis* and *Calamus ornatus* take 1-7 months for completing the germination process. Thc rattan seeds cannot withstand drying out. The whole fruit gives the highest percentage of germination than those of cleaned seeds in *Calamus pseudotenuis* seeds sown in polythene containers filled with a mixture of forest soil and sand, kept in partial shade and watered daily gives better results in germination.

Nursery technique

Cleaned seeds should be kept in moist sawdust in a polythene bag for about 2 weeks till the seeds start germinating. The water is sprinkled over the sawdust when the upper layer gets dry. The process helps to fasten the germination rate. For establishing seedlings, a nursery site should be selected near a perennial water source. The nursery should be partially shaded with a thatch of palm leaves or coir mat as cane seedlings need shade for their initial growth. It is better to sow the germinated seeds separately in polythene containers of 15 x 20 cm size filled with a mixture of forest topsoil and sand in a ratio of 3:1. The polythene containers in the nursery are arranged over a black polythene sheet spread over the soil, so that growing roots do not penetrate the ground. The sawdust with the seeds are transferred to the polythene containers and the seedlings are maintained till the following rainy season *i.e.* June- July.

Alternately, the seeds are sown in a seedbed with a layer of 2-inch-thick sand and farmyard manure mixture. After sowing, the seedbeds should be covered with grass and watered regularly twice a day. Normally it takes 15-35 days and about 85 percent of seeds germinate. After 45-50 days, the small seedlings are carefully prickled out and transplanted in polythene bags filled with 1:1:1 proportion of red earth, sand and farm manure. It has been found that 12 to 14 month old seedlings attain a height of 40-60 cm. For raising plantation of *Calamus tenuis*, sowing the whole fruits within 10 days of collection in the nursery beds gives better results. The seedlings are kept in the nursery for about a year and then planted at a spacing of 5 x 5 m.

b) Vegetative propagation

Usually, the seed is the best source of propagation. However, in rattans, due to uncertainty of viable seeds, vegetative propagation by suckers, rhizomes, auxiliary shoots and stem cutting are the alternative methods for regeneration. Canes are multiplied by suckers and rhizomes. Suckers and rhizomes of 20-25 cm in height can be planted before the onset of monsoon in June-July. Auxiliary shoots developed on the aerial stem can be separated and raised as individual plants. The inflorescence tip of *Calamus pygmaeus* when comes in contact with the soil strikes roots and produces new shoots.

Harvesting

Canes are harvested by cutting mature stems at the base. The soft terminal portions are discarded. The sheaths are removed and then dried in sun or over fire soon after cutting. Canes are extracted between December and May in South India. During the rainy season, they are susceptible to a fungal infection which affects the quality and appearance of canes. Undried canes deteriorate rapidly. The canes are then processed. The silica layer is removed by rubbing the freshly harvested cane over a knife or a sharp piece of bamboo or rubbing it with soapstone. To obtain a fine reddish-brown colour, canes are treated by smoking them over a fire and polishing them with coconut oil. In South Indian states, the extraction of cane is done through departmental agencies or tribal co-operative societies and the harvested produce is brought to the depots. From the depots, it is marketed to different furniture manufacturing units depending on their production capacity and availability of the stock. One major complaint in the cane trade is that the quality of Indian cane products is inferior to that of the canes from the Southeast Asian countries. This is mainly due to a lack of appropriate processing techniques. Canes are generally treated with hot kerosene oil which not only maintains properties like density and strength of natural cane but also improves its colour (ivory white) appearance and durability

Uses of Canes

- Canes find a variety of uses due to their pliability, strength and long length. On account of their remarkable strength, elasticity and length, they are used for various purposes. Most of the canes are very strong and can be used as a substitute for rope cables for suspension bridges and other purposes. They are used as substitute for ropes and cables, walking sticks, polo sticks, umbrella handles, baskets, etc. Canes when split, the strips from the outside are used for making seats, backs of chairs and sofas. Canes are also used in wickerwork and the manufacture of sports goods. The shavings from canes are useful for stuffing, packing, rough cordage and matting.
- The most extensive use of canes is in the furniture industry. Furniture made from canes is in great demand. Chairs, tables, sofa sets, etc., are made from this cane. For making furniture, canes are generally split and the outer shining layer is removed and cut into thin flat pieces, then woven into sheets or backs. Splitting of canes is done usually by hand, machines are also sometimes used. Stems of *Calamus rotang* and *Calamus tenuis* are generally used for furniture making. The thinner canes are generally obtained from *Calamus acanthospathus, Calamus latifolius* and *Calamus viminalis* which are used for making furniture frames, polo wickets, umbrella handles, etc. The refuse from the splitting of the canes is generally used for making rough cordage, fiber mat and variety of other articles.
- The fruits of some species of *Calamus* such as *Calamus rotang* are fleshly, mucilaginous, sweet- bitter and edible. The seeds of *Calamus extensus* and *Calamus erectus* are used as a substitute for betel nut (areca nut). The boiled young tender shoots are eaten or made into curries and also pickled.
- The roots and leaves are used for medicinal purposes. The seeds of *Calamus rheedei* are powdered and applied to ulcers. The roots of *Calamus rotang* is the remedy for dysentery, biliousness and febrifuge and are also used as a tonic. It is used in veterinary as an aperient. The tender leaves of *Calamus travancoricus* are used in dyspepsia, biliousness and as anthelmintic.

Suggested Readings

Albert Spear Hitchcock 2009 *A Text-Book of Grasses,* South Carolina: BiblioLife

Nautiyal S and Kaul AK 2003 *Non Timber Forest Products of India,* Delhi: Jyoti Publishers and Distributors

Rawat SP 2008 *Non Timber Forest Products of India,* South San Francisco: Genetech

Uma Shaanker 2004 *Bamboos and Rattans of the Western Ghats: Population Biology, Socio-economics and Conservation Strategies,* Bangaluru: Ashoka Trust for Research in Ecology and the Environment

Walter Liese, Michael Kohl 2015 *Bamboo: The Plant and its Uses,* New York: Springer

CHAPTER 12
Bidi Leaves

C.N. Hari Prasath and M.P. Divya
Forest College and Research institute, Mettupalayam

Tendu leaves is an important Non-Timber Forest Product (NTFP) that employs millions of tendu leaf collectors each year. To make bidi, tree leaves from the *Diospyros melanoxylon* are utilized as tobacco wrappers. Because of the flexibility with which it can be rolled and its vast availability, *Diospyros melanoxylon* leaf is regarded as the most ideal wrapper. Many other plants' leaves such as *Butea monosperma, Shorea robusta* and others are also used as Bidi wrappers in different sections of the country, but the texture, flavour and workability of *Diospyros* leaves remain unrivaled. *Diospyros melanoxylon* leaves are widely used in the Bidi business because of their large production, pleasant flavour, flexibility, resistance to rot and ability to keep fires burning. Size, thickness of leaves, texture, relative thickness of midrib, and lateral veins are the broad morphological parameters on which leaves are picked and categorized for bidi manufacturing.

In all of the major NTFP producing states, the trade in tendu leaves have been nationalized, which means that only the state government or its approved agents may purchase these leaves from collectors or producers. The major job is bidi rolling, which is quite simple and can be done anywhere at any time. For lakhs of poor rural people, it is a secondary occupation and a source of additional cash. During the off-season, the bidi industry employs the rural populace to collect bidi leaves. The bidi sector plays an important role in rural development. Madhya Pradesh is India's largest producer of tendu leaves. Madhya Pradesh produces over 25 lakh standard bags of tendu leaves each year, accounting for nearly a quarter of the country's total tendu leaves production.

Distribution of bidi leaves

The leaves are obtained from the tendu tree (*Diospyros melanoxylon* Roxb.), which is unique to the Indian subcontinent and belongs to the Ebenaceae family. According to Troup (1921), *Diospyros melanoxylon* is one of the most distinctive trees of India's dry deciduous forests with a distribution that stretches from the Indian plain to the Gangetic plain, Madhya Pradesh, Maharashtra, the western coast up to Malabar and the eastern coast up to Coromandel. In the south, this tree can also be found in the Nilgiris and Serawalli hills. Madhya Pradesh, Orissa, Maharashtra, Andhra Pradesh, Bihar, Rajasthan, Uttar Pradesh, Gujarat, Tamil Nadu and West Bengal have large populations of the species. It can be found growing with *Shorea robusta* and *Tectona grandis* in dry mixed deciduous forests. Sometimes this tree forms pure consociation.

The other species, *Diospyros tomentosa* is closely related to the tendu leaf tree and the leaves are also used for making bidis. Under favourable conditions, the tree attains a height of about 25 m and a girth up to 2.3-2.5 m. The leaves are elliptical, rounded at the base and vary in length from 10 cm to 30 cm, the width of the leaves is 7 to 15 cm. The leaves of the coppice shoots are generally twice as bigger as the normal tree. The leaves of mature trees are shorter, thicker and brittle. The leaves of coppice shoots are prepared for bidi wrapping. *Diospyros melanoxylon*, the main source of bidi leaves, is a medium-sized tree and occurs naturally in forests, wastelands and cultivated lands. It pollards extremely well, pruning the tree is one of the accepted management practices to encourage the growth of new shoots and leaves. The tree develops several root suckers and gets regenerated by creating artificial injuries to its root system. Due to repeated cuttings and hacking, the tree acquires a bushy form.

The species easily adapts to varying conditions of soil and climate and can be raised in arid and semi-arid regions. Generally, the species avoid water logging conditions. The tree is a light demander and is susceptible to frost. It can establish by producing extensive root suckers, which are hardy and immune to grazing by wild animals and cattle. The natural regeneration by seeds and suckers is abundant. Under natural conditions, the seed germinates in the rainy season and seedling production is plentiful. Seedlings develop under moderate shade conditions also but, after reaching the sapling stage, more light is required. The seedlings start producing root suckers, which help in the establishment of seedlings. There is a wide variation in the quality of leaves in different states and regions within the same state. Leaves, large in size, papery in texture, non-brittle and with inconspicuous veins are considered to be superior quality. Such leaves generally fetch five times more price in comparison to inferior quality leaves. The tree can be raised artificially by seed sowing and transplanting of nursery raised seedlings. Germination percent of fresh seeds is about 40, which falls rapidly on storage. Germination starts in about 36 days and completes within two months. Seedlings are transplanted in the field. Direct sowing of seeds in the planting area has been done successfully in several states.

Collection of leaves

The leaves are collected manually from standing trees and bushes. For making bidis, the texture, venation and thickness of the midrib and lateral veins of the leaves are of great importance. The leaves which are soft, pliable, and do not have an excess of pubescence are categorized as good leaves. Leaves are usually collected from the natural growth for making bidis. Plants that are 15 cm in girth are cut near the ground to proliferate the coppice shoots. Coppice shoots yield the best quality leaves and also facilitate collection. Bidi leaves are not collected from the large trees, because the quality of leaves is not good and the collection is also difficult. The leaves are collected from small coppice shoots. In cultivated areas, the plant remains almost at the shrub stage due to repeated cuttings. In February, tendu bushes are frequently coppiced. After around 45 days, the leaves begin to grow and are ready for collection. From the first week of April to the end of June, semi-mature leaves are harvested. The system for collecting and preparing tendu leaves has almost been standardized and the procedure is almost widely followed. During February and March, the tendu plants are clipped. In the tendu leaf production process, tree pruning or bush cutting is a crucial procedure. After roughly 45 days of pruning, the mature leaves are collected. The leaves are removed after they have changed from scarlet to bright green and have a leathery texture. In most cases, the collection begins in the second fortnight of April and continues until the monsoon arrives. Bundles of 50, 70, or 100 leaves are made and tied together with strings or bark fibres. Collectors transport the bundles to the processing locations.

Drying and curing of leaves

The drying of leaves to an optimum moisture condition is very important for their proper utilization. The leaves need to be neither under dried nor over dried because both these conditions cause deterioration in the leaves. The importance of proper leaf drying could be emphasized. If the leaves are not properly dried and moisture is not reduced, the leaves get blackened and affected by molds. Too much moisture causes the leaves to turn dark and moldy, emitting a foul odor and are not suitable to be used in making bidis. Drying the leaves too much makes them brittle, resulting in loss during handling. After collection, leaves are sorted, graded and tied in bundles which are locally known as Gaddies and Pudas. After that, the bundles are turned inside out for 2-3 days. Leaf bundles are stretched out on the ground to dry, with the dorsal sides facing up, for three to four days. In around 8 to 10 days, the drying process is completed. Solar dryers are also used for drying the leaves. In solar leaf dryers, the leaves dry completely within 18-20 hours. There is an additional advantage of solar-dryer that the green colour of the leaf is retained even after drying.

Packaging and storage

Dried leaves are bundled in gunny sacks and stored in godowns until they are sold or used to make bidis. From state to state, the quantity of leaves in a standard bag varies. Before filling the bags, the dried leaves are soaked in water to soften them, then packed tightly into jute bags and exposed to direct sunshine for two days. The bags, which have been packed and cured, can be preserved until they are used to make bidis. When plucking, curing and storing tendu leaves, caution is required. It is a delicate substance and even minor errors or oversights during any of these steps might deteriorate the quality which is inappropriate for preparing bidis.

Marketing of bidi Leaves

Tendu nationalizes trade in all of the producing states. The primary producers of tendu leaves are Madhya Pradesh, Chhattisgarh, Maharashtra, Andhra Pradesh, Orissa and Bihar. Nationalization implies that the government will have a monopoly on trade. Nationalization takes place in different years in different states. The fundamental goals of nationalization are as follows.

- Stopping pilferage in government forests and other lands
- Provide definite value for tendu leaves to growers
- Increase revenue to the state
- Provide adequate wages to labour
- Improve quality and quantity of leaves by regular pruning
- Ensure supply of leaves to small and medium manufacturers of bidis

Trade of tendu leaves in India

In several states, the trade of tendu leaves is under the control of the state government. Orissa was perhaps the first to bring tendu under the control of state government in the year 1963. This was followed by Madhya Pradesh in the year 1965. Today most of the states have taken action to nationalize the trade of tendu leaves. Uttar Pradesh nationalized the trade of tendu leaves in the year 1972. Tendu leaves are one of the most important non-timber forest produce as far as the revenue is concerned. In Madhya Pradesh, tendu leaves generate revenue 2 to 3 times more revenues than from timber and firewood. For organizing the collection, transport, storage and marketing of tendu leaves, the state is generally divided into productive units, based on the production capacity. From a point of convenience, the workable size of units constitutes 2500 standard bags. The quality of tendu leaves varies on a large scale in Madhya Pradesh. Based on the size and other qualities, tendu leaves are generally divided into five quality classes.

Quality Classes

- **Quality 1:** Fairly large size leaves, thin, smoothy textured, the weight of leaves being over 80 kg per standard bag.
- **Quality 2:** Fairly large size leaves, thin and smooth textured, the weight of leaves being 65-80 kg per standard bag.
- **Quality 3:** Medium size leaves, fairly thin and smooth textured, the weight of leaves being 50-66 kg per standard bag.
- **Quality 4:** Small size leaves, thin and smooth in texture, the weight of leaves being 40-50 kg per standard bag.
- **Quality 5:** Small size leaves, thin to moderate textured, weight below 40 kg per standard bag.

The units are notified and the area of each tendu unit is precisely known. Generally, purchasers are appointed for each unit through the tender notice for purchasing the leaves collected by the agents or by the government. If no purchaser is appointed for collection, the collection is generally done by the government department. The purchasers deposit a part price of tendu leaves in advance before taking the delivery of the leaves. Instead of making full payment, the purchasers keep the leaves in the godowns under the control and supervision of the government. The balance amount is generally paid in three installments. The tendu leaves are stored in damp-proof godowns. There is a gradual loss in the quality of tendu leaves during storage in godown. It is, therefore, necessary to take proper steps in controlling the deterioration of tendu leaves under storage.

The entire collection area is separated into many units. The Minor Forest Produce (MFP) Federation at the state level makes cash available to the District Unions for a variety of operations. The District Unions distribute funding to the Primary Societies for procurement and temporary collection centers are set up to gather leaves. In this state, there are about 15,000 collecting centers. At each collecting center, the Primary Society purchases the leaves collected by collectors. At each collecting center, one Forest employee serves as a collector.

Measures for increasing tendu leaf production

There are two important factors in tendu leaf production. The first important factor is the production of leaves in larger quantities in the forests and the second factor is the collection of tendu leaves in larger quantities. Therefore, measures which may lead to increased production and increased collection are important.

a) Increased wages to collectors: The important point in increasing the collection is to have a remunerative price of tendu leaves so that the collectors, who collect tendu leaves are motivated to collect the leaves in larger quantities. In some states, the price of collection of tendu leaves is not good and does not commensurate with the hard work which the collector puts into collecting these leaves. Therefore, proper wages need to be paid to the collectors. The participation of the laborers in the trade of tendu leaves would be the best incentive for increasing the production of tendu leaves from the forests. Such participation has been tried in Madhya Pradesh and it was found to be very effective in increasing the production of tendu leaves. In Madhya Pradesh, the collection of tendu leaves is done by the labourers who are paid remunerative wages. Each collector is a member of the cooperative society and the society is paid additional profit, which the state earns in the trade. The profit is also shared by the collectors in addition to the wages earned by them.

b) Long leases: Some people think that if the units are sold for longer periods *i.e.* 3-4 years, the purchaser will take interest in improving the production because the purchaser will be interested in undertaking operations such as coppicing, pruning and carrying out other operations which will increase the tendu leaf production in that unit. The sale of units for one year does not provide an incentive to the purchasers for such operations.

c) Adjusting the period of coppicing: The period of coppicing and pruning may be adjusted in such a way that the damage to new leaves on account of insects, pests and fungi is minimized. Several insect pests, defoliators and diseases are reported to occur on tendu leaves. It is observed that when light rains are received while the flush of new leaves is sprouting, fungi and insect-pest attack on the leaves is more. These attacks make leaves unsuitable for bidi making. The defoliators that damage tendu leaves are *Hypocala rostrata, Lamida carbonifera*, etc. Extensive damage is caused by the gall fly (*Psylla obsolete*), which causes blisters on the leaf surface. Leaf pox disease causes extensive damage to leaves and makes them unsuitable for bidi. The fungi which affect tendu leaves are *Lenzites repanda, Stenrum lobatum,* and *Thalephora* spp., etc. Due to the attack of these fungi, tendu leaves develop circular holes and the leaf quality deteriorates. These insect-pest attacks can be minimized by readjusting the pruning and using some chemicals for controlling these insect pests.

d) Meteorological conditions: The meteorological data need to be collected and the work of tendu leaf collection should be adjusted depending upon the weather conditions. The period of new leafing and leaf maturation also requires attention in such adjustments. The period of collection of tendu leaves is short and generally varies from 15-20 days. There is a particular stage in which leaves are to be collected. The new leafing should be carefully watched and the collection of leaves should start at a particular time. Similarly, when pre-monsoon showers start, the collection can be extended by resorting to the early collection and taking the advantage of full period before the monsoon showers. The quantity of tendu leaf collection can be increased by proper monitoring of weather conditions and new leafing.

e) Improving storage and transport facilities: The present conditions of storage of leaves and facilities for transport are poor which work as a constraint in tendu leaf collection. The proper storage and transport facilities at different places of leaf collection will help in the collection of tendu leaves and also minimize the losses.

f) Plantation management: Regeneration of the species by seed sowing and planting needs to be a standardized and efficient method of the plantation, management need to be developed for proper collection of tendu leaves. The plantation of tendu leaves in the form of tea gardens needs to be worked out for maximizing production. In large trees, if the roots are slightly exposed and damaged, several root suckers are produced. These plants from root suckers can yield additional leaves.

g) Increasing the period of collection: The tendu leaves are generally collected once a year. Efforts may be made to collect the tendu leaves more than once a year. It has been found that harvest can be taken from tendu crop by adjusting the period of coppicing. If the crop is coppiced in the first week of January, the harvesting period of leaves may be the first week of March. If the coppicing is done during the third week of March, then leaves may be ready for harvest in the last week of April. Similarly, if the plants are pruned in the first week of June at the onset of monsoon. Thus, the tendu leaf collection period can be extended from March to June by adjusting the coppicing period.

Disposal of tendu leaves

The leaves which are collected by the state governments or their agents are generally kept in the godown. These leaves are sold to different purchasers by the method of auction or tenders. Efforts are made to sell and dispose of leaves as soon as possible. These leaves cannot be stored in godowns for a longer period because leaves degrade rapidly. The value of stored leaves falls rapidly and after a year or two, these become completely useless.

Utilization of leaves

Tendu leaves collected are sold to the traders or manufacturers of bidi and are utilized for bidi rolling. Since many states have controlled the tendu leaves trade, the state governments play a major role in the trade of these leaves. Generally, 15-50 percent of leaves are locally used and the remaining are exported to the different states and even to other countries. The main bidi manufacturing units are established in Madhya Pradesh, Maharashtra, Kerala, Tamil Nadu, Andhra Pradesh, etc. Madhya Pradesh has the major industries at Jabalpur, Sagar, and Bhopal.

The bidi rolling is practiced as a cottage industry and mostly ladies are employed for this work. Manufacturers issue them tobacco and leaves for rolling bidis. Bidi rolling is a simple process, it does not require any special training. Before rolling bidi, the leaves are kept covered with moist gunny cloth for about 12 hours, or overnight. The leaves are cut in rectangular shape according to the size of the bidi to be rolled. The cut leaf is then rolled by hand into a conical shape stuffing tobacco mixture in it. The other end is tied with a small piece of thread. The standard sizes of bidi are 5 cm, 6.25 cm and 7.25 cm in length.

The quantity of tobacco used per 1000 bidis ranges from 225 to 350 grams depending upon the size of bidis to be rolled. One kg of the leaf can make about 750 to 1000 bidis. Mixtures of tobacco produced in Gujarat are preferred for bidi making. The composition of the tobacco mixture is the trade secret of the bidi manufacturer. After bidis are made, they are dried and excess moisture is evaporated. Drying of bidi leaves is generally done in ovens specially set up for this purpose.

Other bidi leaves

Generally, *Diospyros melanoxylon* and *Diospyros tomentosa,* two species of *Diospyros* are used for rolling bidi in different parts of the country. The following species are also used in some parts of the country for bidi rolling.

Kuda leaves: The leaves obtained from *Holarrhena antidysenterica* are commercially called kuda leaves. Kuda leaves are also used for making bidis. The use of these leaves for bidi making is restricted in some parts of Maharashtra, Goa and Andhra Pradesh. Plucking of these leaves is undertaken manually on the slopes of Western Ghats in the Ratnagiri district. The main center of production of kuda leaf bidi is Ratnagiri district (Maharashtra) and Goa. Plucking of kuda leaves is done in October and November after the leaves are fully matured before they turn yellow. Overmature leaves which starts turning yellow are unfit for bidi rolling. Leaves from coppice shoots and root suckers are considered better. The leaves from the kuda tree are collected and dried in the same way as tendu leaves. The market rate of these leaves is cheaper than the tendu leaves. These bidi leaves are commonly found in the market of Bombay. More recently, these bidis are giving way to tendu leaf bidis.

Bauhinia leaves: Next to tendu leaves, the leaves of *Bauhinia racemosa* are used for bidi making particularly in Maharashtra, Andhra Pradesh and some parts of Karnataka. The leaves are collected twice a year, once in April- May and then in September- October. The leaves are smaller and thicker and possess better texture and smoking quality. These leaves fetch a better price than the leaves collected during September - October. The leaves are dried in the same way as tendu leaves. The drying is generally done by making the garland of leaves and hanging them under shade. Garlands made of green leaves are dried and delivered to the purchasers. The leaves from the Thane area (Maharashtra) are considered to be better in quality. The leaves have two lobes and can be separated easily for making bidis. It does not require cutting for bidi-making. Leaves are moistened before rolling bidis. Generally, 400-800 bidis are made out of one kilogram of dried leaves. The burning rate of these leaves is found to be one cm of bidi per 240-280 seconds. The smoking quality of these bidis is considered better than tendu bidis by some smokers.

Pena's leaf: Fresh leaves of jack fruit (*Artocarpus integrifolia*) are also used as a bidi wrapper. This type of bidi is not available in the market. The leaf is wrapped into bidi by the smoker himself or herself. This type of bidi is popular among fisherwomen of Goa. The shape of this bidi is like an ice cream cone and it is not tied with thread but held together by a palm veinlet. The lit end of the bidi is put in the mouth and smoke is inhaled.

Banana leaf: The leaves of banana (*Musa paradisica*) dried on the plant itself are also used as bidi wrapper. This type of bidi is also made by the smoker himself. The size of the bidi is like a cigarette.

Suggested Readings

Dhiman KR Bawa R Raina KK and Dohroo NP 2013 *Non Timber Forest Products: Conservation Utilization and Management*, Dehradun: BSMPS Publishers

Krishna Murthy T 2010 *Minor Forest Products of India (Non-Timber Forest Products of India),* Hyderabad: BS Publications

Rajeev Joshi and Ramesh Chhetri 2021 *A Textbook of Non-Timber Forest Products*, Bhotahity, Kathmandu Nepal: Heritage Publishers and Distributors Pvt. Ltd

Tribhuwan Mehta 2008 *A Handbook of Forest Utilization,* Delhi: Periodical Expert Book Agency

Uday Kumar V and M M Rehman 2005 Tobacco Tendu Leaf and Beedi Workers in India: Problems and Prospects, Published by Shipra, Delhi

Vinod Kumar 2001 Tendu (*Diospyrus* spp.). Daya Publishing House, Delhi

CHAPTER 13
Dona Making

V. Priyanka and B. Vinothini
Forest College and Research institute, Mettupalayam

The disposable plates, cups, bowls, tumblers, spoons, bags, covers, sheets, films are made from plastics such as polythene, polypropylene, polyester, polyethylene terephthalate, polystyrene, polycarbonate, epoxy resins, polysulfone, polyvinyl chloride, polyvinylidene chloride and melamine formaldehyde. These plastics can release toxic substances including bisphenol A, melamine, vinyl chloride, phthalates, etc. into the food during usage. Notably, most of these plastics are produced from nonrenewable fossil fuels such as petroleum or natural gas, which are going to be exhausted with increasing demand and show definitive impact on human health and environment. It is alarming that India generates around 26000 tonnes of plastic waste a day and each Indian consumes 11 kg of plastic annually.

The impact of disposable plasticware utilization in humanbeing day to day has led to a search for alternate renewable resource, *i.e.*, use of plant leaves as dining plates and food wraps, a traditional practice in India. The long-standing tradition has its own cultural, religious, medicinal, socioeconomic importance in India. The leaves are one of the Non-Timber's Forest Produce (NTFP) and collected from the forests by the tribal people of India. The plates and cups made out of leaves are known as patravali, pattal, vistari, vistaraku and dona in various Indian languages. The leaf plates are environmental friendly, biodegradable and amenable for longer duration storage and can be easily disposed off. They are economical and don't require cleaning with phosphate rich soaps and detergents. The leaves exhibit significant antibacterial and antifungal properties against various bacteria and fungi, thus protect the human from the environmental and food borne pathogens (Fig. 13.1).

Fig. 13.1 Dona prodcuts using tree leaves

Suitable tree species for dona making

In olden days, the leaves are extensively used for packing breakfast, meals and groceries and the practice is continued in villages even now. Varieties of plant leaves employed as plates, wraps and packing material in India. Plant and tree leaves belonging to a variety of families of the plant kingdom are used as single leaf plates, stitched dining leaf plates, food wraps and food packing materials.

Shorea robusta

It is known as sal, is a large, deciduous, highly regenerating, high timber valued tree belongs to the family Dipterocarpaceae. It has religious importance for Hindus, Jains and Buddhists. It is a native tree of Indian subcontinent and distributed in Odisha, West Bengal, Bihar, Jharkhand, Madhya Pradesh, Chhattisgarh, Haryana, Assam, Tripura, Andhra Pradesh and Himachal Pradesh states of India. The leaves are 10-25 x 5-15 cm size and used in Ayurveda and Unani systems of medicine. The fresh leaves are used for serving small snacks and the sun dried leaves are stitched together using grass stem sticks or sewing machine to produce leaf plates (khali) and leaf bowls (dona). The leaf plates are used for serving meals during domestic and ceremonial occasions. The leaf cups are routinely used for serving prasadam to the devotees in temples. As the sal forest of Odisha occupies around 30% of total sal forests in India, it is one of the revenue generating activities for the landless, marginal and forest dependent tribal people living in the Sundargarh, Mayurbhanj, Kendujhar, Kandhmal, Debagarh, Nayagarh, Balasore and Sambalpur districts of Odisha. Due to the absence of legal restrictions, round the year leaf availability and occurrence of traditional knowledge in leaf plate making, the local people adopted this occupation as a source of income.

Phanera vahlii

It is a giant, wild, woody, perennial climber belongs to the family Caesalpiniaceae, a native tree of Indian subcontinent. It is widely distributed in the forest areas of Odisha, Andhra Pradesh, West Bengal, Telangana, Karnataka, Tamil Nadu, Maharashtra, Bihar, Uttar Pradesh, Madhya Pradesh, Rajasthan, Punjab, Haryana, Uttarakhand, Sikkim, Assam, Arunachal Pradesh, Nagaland, Manipur, Mizoram, Meghalaya, Himachal Pradesh and Jammu and Kashmir. The leaves are bilobed, 10–46 cm long, rich in various quercetin, flavonoids and known to exhibit antibacterial, antimicrobial, antioxidant, anti-inflammatory and antidiabetic effects. The shade dried leaves in 4-5 numbers are hand and machine stitched into round plates using grass stem sticks and thread respectively. These leaf plates are extensively used as meal plates in religious ceremonies, festivals, community feasts and also in occasions where food is served for a large number of people. They are also made into cups and bowls for serving the liquid foods. In Kurnool district of Andhra Pradesh state, the hand woven cups are often used for serving honey along with sajja rotte by the Chenchu tribes of Nallamala forest at the roadside eateries.

In Odisha state, the plate making activity with siali leaves is one of the most important socioeconomic activities for the people in the forest villages. It is especially a means of subsistence for the rural women in tribal dominated districts of Odisha and Paderu forest division of Visakhapatnam district of Andhra Pradesh. In comparison with sal, siali leaves are the preferred raw material due to the large size and durability of the leaf. As the leaves are critical source of the income, the tribal people protect and nurture the siali climbers. Some of the Non-Governmental Organizations (NGOs) such as Pradan are collating the makers as self-help groups and forming cooperative societies for increasing the income of the local village tribal people by ensuring higher profit margins. It is done by providing training, technology, market linkages with wholesale buyers and upgrading product quality. It is estimated that sal and siali leaf trade in Odisha is around Rs.400 crores every year.

Butea monosperma

It is known as flame of the forest and is a medium sized, deciduous, drought and frost resistant tree, adapted to grow extensively in poor soils of open plains, grasslands and wastelands in the dry seasons. It belongs to the family Fabaceae, a native tree of tropical and subtropical climate of Indian subcontinent and Southeast Asian countries. In India, it is abundant in the states of Jharkhand, Andhra Pradesh, Telangana, West Bengal, Maharashtra, Kerala and Punjab. The leaves are trifoliate, pinnate and 10 -15 cm long. In the states of Odisha, Maharashtra, Andhra Pradesh and Telangana, the plates (vistari) and cups (dona) made of dried leaves are widely used for serving food and meals for people during rural feasts and also offering naivedyam to various gods and distribution of prasadam to deities during festival and religious ceremonies. It is believed that by eating the food served on moduga leaves keeps the person away from liver disorders, skin diseases and cancer. It is one of the main host trees for the lac insect (*Laccifer lacca*) and pruned trees are the ideal candidates for lac cultivation. Hence, the leaf plucking is done in an environmental friendly process without causing damage to the tree.

Terminalia catappa

The Indian almond is a medium sized, tropical tree that belongs to the family of Combretaceae. The ovoid shaped leaves are 15-25 cm long and abundant in various polyphenols, flavonoids, tannins, saponins, and phytosterols. The fresh leaves are hand stitched into plates and cups and used for serving food.

Tectona grandis

The teak is a large, tropical, hardwood, deciduous tree that belongs to the family Lamiaceae and naturally distributed in the Indian subcontinent. It is abundantly distributed in the states of Tamil Nadu, Karnataka, Kerala, Odisha, Madhya Pradesh,

Jharkhand, Maharashtra, Rajasthan, Bihar and Telangana. The leaves are large, rough, broadly ovate to elliptic and sized around 8-60 x 15-30 cm. The teak leaf cups are used for serving liquid food items such as soups and dal.

Pterospermum acerifolium

It is a tropical, ever green, ornamental, landscaping tree that belongs to the family Sterculiaceae, a native tree of Indian origin distributed in the states of Himachal Pradesh, Chandigarh, Delhi, Madhya Pradesh, Odisha, West Bengal, Sikkim, Meghalaya and Manipur. The leaves are simple, very broad, palmately ribbed and sized around 40 x 40 cm. The leaves are used for serving food, making disposable plates, packaging and storing material and fodder. They are also woven into dinner plates and bowls either by stitching with twigs or molding. Because of its large sized leaf blade, it earned the name dinner plate tree.

Dona Manufacturing Process

Siali (*Phanera vahlii*) is a creeper with large leaves that typically grows around Sal trees and is commonly found in the forests of Odisha and parts of Andhra Pradesh, Chattisgarh and Jharkhand. On account of its size, texture and durability, siali leaves are very good source of raw material for the production of disposable plates and bowls. Six different products are made from Siali leaves. Loose leaves for dry snacks; stitched and moulded plates are used in functions. There are 2 grades of siali leaves viz., i) Clean, dry and big leaves and ii) Damaged or slightly black leaves that are normally sold in pan and flower shops.

The size of the leaves varies from season to season. Small sized leaves appear during summer months with the size varying from 9 to 14 inches. In the rainy season, however, sizes could go as long as 18 to 24 inches. In the summer, the texture of the leaves remain fine and shiny and there is less damage to leaves as they dry quickly. Generally, the green, thick and large sized leaves are preferred. Tribal women travel deep into the forests to collect these leaves, which are then stitched together with bamboo splits to make disposable plates. These products are the familiar sight as the bowls on which temple prasad or pani puri is served.

Leaf collection

Women travel deep into the forest to collect the leaves. The leaf is then bundled into chakkis, where each chakki consists of 80 to 100 leaves. Most people who collect the leaves also stitch them, since it provides them with an additional source of income. If they are to sell the chakkis, they would get as little as Rs 10 - 16 for a chakki.

Stitching

Leaves are stitched together with bamboo splits and are graded into raw-stitched, medium stitched and fine stitched based on the number of stitches holding the leaves together.

Rough Stitched: Rough stitched plates have two to three stitches between each leaf and in no particular orientation. They look patchy and mostly the poor quality leaves are used in rough stitched plates.

Medium Stitched: Medium stitched plates have five to six stitches between two leaves, where the stitches also follow an orientation to make it neat. Medium stitched plates are also leak proof, if the leaves have holes they are patched.

Fine / Close Stitched: Fine stitched plates have twelve to fourteen stitches between two leaves, where the stitches follow a close orientation to make it look pleasing to the eye. These plates are also leak proof as the holes or gaps are patched.

Machine Stitched: Machine stitched plates are made with a sewing machine. These machines are given by the cooperative to its member for a fee of Rs 40 per month. The quality of this stitched plate is really good as the stitches are finely oriented, and is completely leak proof.

Cutting

The leaves are stacked and manually cut to make the plates look perfectly round in shape and thus more appealing. A majority of the cut leaves are directly packaged and sold to the main metro markets in Andhra Pradesh and Maharashtra. Only a small portion moves to the next level of value addition, which is moulding.

Moulding

Moulding into buffet plates is a product that serves only an elite market. Family weddings, important functions and big hotels use moulded plates. The cut leaves are made to have a base of cardboard and a thin plastic sheet is inserted in between to make it leak proof. This arrangement is put into a moulding machine which will then produce the buffet plates. These plates look fancy and are the most expensive of all Siali plates.

Packaging

The finer the packaging, the higher the price fetched per chakki. The cut leaves or moulded leaves are put into plastic bags in bunches of 20 or 40. These are transparent plastic covers. To make it look fancy, a round gold sheet as per the exact size of the plate is inserted at the top and the bottom of the bunch to make it look good from both sides. These sheets can come in many colors, although gold is the most popular as it looks very attractive when it is packed. Then the bags are sealed. Then they are stacked in bunches of 80 or 100 @ 5 bags of 20 or two bags of 40. The bunch is then wrapped in a cane mesh and tied with jute rope. The whole package looks attractive, as it shows gold on both sides and a thick cane mesh, holding it all together.

Branding

The packages are branded before sealing. After the gold paper is added on both sides, the worker also inserts a small square sheet of white paper on which the brand of the plate is printed along with a logo. In comparison with plastic plates, the leaf plates and cups offer many advantages such as renewability, biodegradability, non-toxicity and antioxidant abundance in addition to their religious, medicinal, socio-economic importance in Indian culture. It is evident that there is a large scope in leaf plate technology in terms of raw material quality, design improvement, product range, production process and marketing. In addition to national market, a huge demand exists in international market for disposable and biodegradable leaf plates.

Suggested Readings

Chouhan R and S Saklani 2013 *Bauhinia vahlii*: a plant to be explored. International Reaserch Journal of Pharmacy, 4: pp 5–9.

Rai A A K Singh V C Pandey N Ghosal and N Singh 2016 The importance of *Butea monosperma* for the restoration of degraded lands. Ecological Engineering, 97: pp 619–623.

Sahu M C and R N Padhy 2013 In vitro antibacterial potency of *Butea monosperma* Lam. against 12 clinically isolated multidrug resistant bacteria. Asian Pacific Journal of Tropical Disease, 3: pp 217–226.

Sahu P 2015 Sal and siali leaves sustaining tribal Orissa. Down To Earth.

Shriya M 2009 A document on the Siali leaf-plate making activity of Kandhamal district, Orissa. Pradan. Accessed 27th August 2018.

CHAPTER 14
Animal Products

M.P. Divya and S. Krishnamoorthi
Tamil Nadu Agricultural University, Coimbatore

Animals always played an important in the life of human beings. Man domesticated certain animals are meeting various needs. Animals found in the forest attract people due to several reasons. Animals and birds attract people with sweet perching sounds, dance, games and their habitat. Small animals and insects yield some products valuable for mankind.

1. BEE KEEPING

Beekeeping is an industry concerned with the care and development of bee colonies in such a manner that the bees store more honey than they need for the food and the surplus is made available for human's consumption. Honey bees are social insects that produce honey and bee wax. The practice of domesticating bees was first adopted by western countries. History reveals that beekeeping in India also has a long-chequered record. In ancient India, man domesticated bees for honey and other products. During those days, honey was used in sacramental rites and medicine. Almost all the known species of honey bees are found distributed in India. Earlier for the extraction of honey from the comb, crude, unhygienic and unscientific methods were employed. Beekeeping when tried commercially on large scale is termed as Apiculture. The bee colonies are hived in special boxes of specified dimensions in a place is known as an Apiary. In India, for the last forty years attempts have been made to run apiaries on modern lines. Honey bees belong to the family Apidae under the order Hymenoptera. They are social insects, which live in colonies. A beehive is a colony of three different types of bees, one queen or mother bee, a large number of sterile females classed as workers and in some seasons, male bees or drones. Generally, a normal colony has one queen, twenty-five thousand workers and a few hundred drones in a comb. A single comb with cells is used as cradles for young ones and as a store for honey.

Important bee species

Apies dorsata: It is commonly called a rock bee and is one of the largest Indian species distributed throughout India up to an altitude of 2700 m. It is a good honey gathered. Each colony of this species builds a huge comb measuring about 1.25 m in length and about 0.75 m in width, in the open or hanging rocks, under branches of large trees and under the ceilings of buildings. A single comb yields about 36 kg of honey and about 1.5 kg of beewax. Rock bees are irritable and vicious and migrate from place to place in different seasons, so these cannot be kept in hives. In India, this species is the principal source of honey and beewax. Honey collector's burn or smoke away from the bees and cut their comb. To obtain honey, the combs are squeezed and melted for beewax.

Apies indica: It is commonly called as Indian hive bee, found almost throughout India. The bee lives in building combs in covered situations such as the cavities in tree trunks, rocks or walls. It also occupies unused boxes, packing cases and protected niches in the buildings. It can be domesticated in artificial hives which construct multiple parallel combs with an average honey yield of 6-8 kg per colony per year. These bees are larger than *Apis florea* but smaller than *Apis mellifera*. They are more prone to swarming and absconding. They are native to India.

Apies florea: It is commonly called as the little bee, found in Assam, Bengal, Tamil Nadu and the Central parts of India. This species is not much commercial importance; only a little amount of honey is obtained from a single comb.

Apis mellifera: It is called as European bee or Italian bee. They are also similar in habits to Indian bees which build parallel combs. They are bigger than all other honeybees except *Apis dorsata*. The average production per colony is 25-40 kg. They have been imported from European countries (Italy). They are less prone to swarming and absconding.

Dammar bee: Besides true honey bees, two species of stingless or dammar bees, viz. *Melipona* and *Trigona* occur in India in abundance. These bees are much smaller than the true honey bees and build irregular combs of wax and resinous substances in crevices and hollow tree trunks. The bee is dark-coloured, small-sized and stingless. They live in small combs generally made in the cavities of tree trunks or walls. The mouth of the comb protrudes outwards and is in the shape of funnel. The comb of these bees is made of earth, wax and resinous material. But the honey yield per hive per year is only 100 gm and it has great medicinal values. Stingless bees have importance in the pollination of various food crops. They bite their enemies or intruders. It can be domesticated.

Modern beekeeping in India started in the year 1789, with the invention of the artificial hive. The commonly used model of an artificial hive consists of a rectangular wooden box open from both ends. The top of the box is covered with a movable roof. Inside the box, combs are supported by a few movable frames. The box is made stationary on a broad wooden piece that functions as a floor and as an alighting board. To give freedom of movement to the bees in the box, the important requirements are the movability of the frames and the arrangement of space between the frames. The other requirements are i) artificial comb foundation, ii) queen excluder, iii) entrance guard, iv) smoker and v) honey extractor. For making beekeeping successful, a suitable breed of honey bees can be domesticated, the only species found suitable is the Indian hive-bee.

Bee Products

The important bee products viz., honey, bee wax, bee venom, propolis, royal jelly and pollen are described below (Fig. 14.1).

a) Honey

Honey is a sweet, viscous food produced by honey bees and other types of bees. Bees make honey via regurgitating, enzymatic activity and water evaporation from the sugary secretions of plants (floral nectar) or secretions of other insects (honey dew). Honey bees keep their honey in wax structures known as honeycombs, whilst stingless bees keep their honey in wax and resin pots. Because of its widespread commercial production and human use, honey produced by honey bees (genus *Apis*) is the most well known.

Honey is extensively used in many preparations in Indian medicine. Honey, especially lotus honey is useful in eye troubles. Being demulcent gives relief from sore throat, cough and colds and constipation. It is used as an antiseptic and is applied to wounds and burns. It is used as an ingredient in cough mixtures, expectorants and sedatives. It is largely used in preparing soft bread and cakes and also in candies and soft drinks. It is used in the preparation of creams and lotions. The solution of honey has a good standing property and does not freeze under ordinary conditions. It is used as an antifreeze solution in automobile radiators.

Bees store honey in the cells of the honeycomb and remove the extra moisture content by a process of natural ripening and seal the cells filled with honey using wax. When about 75 percent of the cells are filled with honey and sealed, the combs can be taken off for the extraction of honey. The bees are driven off and the cells are uncapped by cutting with a hot knife. Honey is separated using the honey extractor. The excessive moisture is removed by heating at 60-65°C for about half an hour. Raw honey extracted from hives is filtered through a sieve or through a muslin cloth and allowed to ripen in the tanks for about 24 h. The colour of the honey varies from straw yellow to reddish-brown and even black. It also varies in flavour depending upon the pasturage. The blending of honey is graded according to its colour, flavour, aroma and density. Honey's composition varies depending on its geographical location, proximity to a flower garden or crop area and the type of honeybee that makes it.

Honey

Beewax

Honey comb

Propolis

Royal jelly

Fig. 14.1 Bee products

Harvesting and extraction technique

Honey harvesting processes differ depending on the type of beehive. It involves multiple phases in modern bee colonies, including colony management, shaking or brushing the bees, transferring the bee colony and the extraction process. During the extraction procedure, a frame with at least 2/3 of ripe honey is chosen. In cold and humid weather, the frame is kept in a heated chamber to facilitate honey extraction. The room's warm temperature is maintained with the use of a room heater, firewood, or coal. The comb will get softer and more likely to break as the temperature rises. As a result, the temperature of the extraction should not exceed 30°C. The wax capping in the honey comb is first removed on both sides with a sharp knife. It is then placed in the honey extractor drum once the combs have been opened. The honey extractor drum has a bicycle-style pedal. The honey is extracted from the comb by centrifugation and collected in the extractor drum.

Processing

Honey processing indirectly entails heating honey. Batch pasteurization at 63°C for 5-6 min is commonly used during this process. Excessive heat will degrade the nutritive value of the food. Another technique of heat treatment is to leave honey out in the sun for 30 min. The honey is then stored in a cylindrical container for filtering and purification. Mineral and metallic particles will settle whereas foreign elements, wax particles, bug fragments and trash will float to the surface. In addition to settling, straining is utilized. Simple metallic screens coated with a thin nylon mesh or a nylon sack filter submerged in a tall, narrow tank can be used as strainers. The smallest mesh size frequently utilized is 0.1-0.2 mm in diameter. The straining temperature should be at 30°C.

Packaging

Honey and honey containing products require their own set of packaging materials. Glass or stainless steel storage containers for liquid or crystallized honey should be airtight and coated with food safe plastic, paint or beewax. Glass is the most common packing medium for the pure honey sale followed by plastic and for large quantities, metal containers coated with materials suitable for interaction with acidic foods. The honey must be easily removed from the container. The label, container form and substance (other material) should all be chosen in this manner. Recycled glass bottles may be appropriate if they can be cleaned adequately and a cork type seal can be provided. However, recyclability, disposability and environmental friendly packaging materials should be used for packaging.

Storage

Honey is a stable product in the concept that it is not damaged by the bacteria and fungus that normally cause food to spoil. However, there are various elements that cause honey to deteriorate, including microorganisms, acidity, heat and sunlight. Unprocessed honey, whether crystallized or liquid, is still vulnerable to fermentation. As a result, storage conditions must avoid fermentation by either storing at a low temperature or preventing further moisture adsorption. Faster degradation is caused by a change in pH (more moisture and low pH occurs). Honey quality can be affected by heat and sunlight. Glucose oxidase is destroyed by ultraviolet radiation. Only cold storage below 5°C is suitable for preventing crystallization, melting of crystallized honey and fermentation at the same time. However, such storage is costly. Only retain liquid honey above 25°C to avoid crystallization. For the storage of liquid or crystalline honey, a temperature of 20°C is suggested as a solution. As a result, the room should be kept at a temperature of around 20°C with a relative humidity of less than 65 per cent. Honey stored at temperature above 25°C loses quality over time as a result of chemical and enzymatic changes. Preservatives will be used less or not at all if proper storage and packaging are combined with speedy marketing and consumption.

Composition of fully ripened honey

- **Pigments:** Carotene, chlorophyll and xanthophyll are the important pigments present in honey.
- **Minerals:** Potassium, calcium, phosphorus, sodium, magnesium, manganese, copper, sulphur, silica and iron are the minerals present in honey.
- **Vitamins:** Vitamin B_1 (thiamine), B_2 (riboflavin), nicotinic acid, vitamin K, folic acid, ascorbic acid and pantothenic acid are the vitamins present in honey.

Honey, however, is high in minerals, proteins and amino acids, vitamins, enzymes and around 181 other minor ingredients in addition to sugars. Honey, royal jelly (which contains protein, sugar, vitamin B, sterols and other nutrients) and pollen, which has been widely used as a nutritive food as well as for medicinal and pharmacological purposes since ancient times.

Physical properties of honey

- Honey is hygroscopic. If exposed to air it absorbs moisture
- Honey is a viscous fluid.
- Heating of honey reduces viscosity

- Specific gravity of pure honey is 1.35 to 1.44 gcc^{-1}
- Refractive index of honey, helps to find moisture content measured using refractometer.

Quality test

Purity test for honey: Measure specific gravity of honey using hydrometer. If the specific gravity is between 1.25 to 1.44, it is pure honey.

Aroma and flavour of honey: It is acquired from the nectar of the flower. It is lost if heated or exposed to air for a long time.

Colour of honey: Depends on the nectar of the flower and the plant species. Darker honey has stronger flavour. Lighter honey has more pleasant smell.

Crystallization or granulation of honey: This is a natural property of honey particularly at low temperature. Dextrose is present in honey granulate and settle down. Levulose and water remain at top which is more prone to fermentation. High ratio of levulose: dextrose (L/D) results in less granulation. High ratio of dextrose: water (d/w) results in more granulation.

Uses of honey

Honey has value as a food, as a medicine, as a cash crop for both domestic and export markets and as an important part of some cultural traditions.

As a food: Honey is valued everywhere as a sweet and tasty food. At times of food shortage, it is a useful carbohydrate source that contains trace elements and adds nutritional diversity to poor diets. Honey often has an important place in traditional food preparation.

As a medicine / tonic: In many parts of the world, honey is used as a medicine or tonic and as a special treat for children. Modern medicine is increasingly using honey for a variety of treatments.

As a cash crop: Fresh local honey is always more highly valued than imported honey. Many beekeepers sell their product directly to consumers. Honey is often used as a barter commodity in villages, especially in remote areas or areas isolated by war or sanctions. Honey is a stable commodity with a long shelf life. If harvested carefully, it will remain wholesome for many years.

As an export crop: As standards of living rise, honey consumption increases. Most industrialized countries import honey to meet demand. The countries with the highest honey exports are Mexico, China and Argentina.

b) Beewax

Beewax is the material that bees use to build their nests. It is produced by young honeybees that secrete it as a liquid from special wax glands. Worker bees secrete wax when they are 14 to 18 days old. On contact with air, the wax hardens and forms scales, which appear as small flakes of wax on the underside of the bee. About one million wax scales make 1 kg of wax. Bees use the wax to build the well known hexagonal cells that make up their comb, a very strong and efficient structure. Bees use the comb cells to store honey and pollen; the queen lays her eggs in them and young bees develop in them. Beewax is produced by all species of honeybees, although the waxes produced by different species have slightly different chemical and physical properties.

The main source of beewax in India is *Apies dorsata* hives. Capping of hives of domesticated bees collected during honey extraction and discarded combs in apiaries are other sources of beewax. Wax is extracted by heat or pressure when combs are boiled with water and rises to the top and solidifies on cooling. Large apiaries use mechanical devices for obtaining beewax. Crude wax is melted by means of steam and freed from solid impurities, it is then cooled and when the molten material solidifies it is ready for marketing. It is a lemon yellow to orange in color. Good quality, unadulterated beewax is clear in appearance. It is used for various purposes. Before using it for making articles, the wax is generally bleached.

Beewax (unrefined) is a non-crystalline solid, with a honey-like aroma, tasteless, brittle and soft to touch. It is used in making foundations for artificial hives, furniture, floor polishes, electrical insulators for grafting, scaling and sewing waxes. A solution of beewax with castor oil is used as a protection cover against mustard gas. Refined beewax is used for making candles, cosmetics, ointments, suppository bases, plasters and surgical dressings.

Composition and properties

It consists of alcohol's and fatty acids @70 to 74 per cent; free acids @13 to 15 per cent; saturated hydrocarbons@ 12 to 15 per cent; vitamin A@ 40961U; specific gravity @ 0.95 and melting point of 65°C.

Processing

Beewax is obtained from the cappings collected during honey extraction. Wax is obtained from old combs that are unfit for use and from combs damaged during honey extraction. Best grade wax is obtained from cappings where the recovery percent is higher. In India, major proportion of wax is from combs of *Apis dorsata.*

Uses

- Used by candle industry, preparing comb foundation sheets, cosmetics like cold creams, lipsticks and rouges
- Used in pharmaceutical and perfume industries (ointments, capsules, pill coating and deodorants)
- Used for preparing shoe polish, furniture, etc. for water proofing
- Used in adhesives, chewing gums and inks, etc.
- Used in the manufacture of cosmetics like beauty lotions and creams, lipsticks, ointments and pomades and polishes for boots, floor and furniture, lubricants, paint and varnishes, inks, electrical insulating apparatus and candles.

c) Bee Venom

Sting of worker bee is attached to a poison sac where venom is stored. Newly emerged bee is unable to sting because she cannot insert the sting which is not fully chitinized and also little amount of venom is stored in the venom sac. A bee, when two weeks old has maximum venom in its poison sac.

Properties

Bee venom contains histamine, apamine, acithinase, hydrochloric acid, formic acid, orthophosphoric acid, sulphur, calcium, copper and magnesium sulphate.

Production

Bee venom is commercially obtained by the use of electric shock. An electric current is passed through copper wires at 12 volts. The bees get shock, irritated and release venom by inserting the sting into a thin nylon cloth below the copper wires. Venom is deposited on a glass plate placed below the nylon sheet. The venom on drying is scrapped from the glass plate.

Uses

- Rheumatism can be cured by apitherapy, where bees are made to sting the patient
- Venom can be used as sub-cutaneous injection for treating rheumatism
- Ointment made by mixing apitoxin, vasaline and salicylic acid (1:10:1) can be applied on affected areas. It has stimulating effect on heart muscles and decreases cholesterol level and lowers blood pressure.

- It can cure neurosis, endoarteriosis, endoarthritis and neuraglia.
- Antihistamine creams or injections are used as anti-allergents.

d) Propolis

Propolis gathered by bees from resinous exudes of tree. In the bee colony, propolis is used for sticking frames, sealing cracks and crevices but it is a contaminant of comb wax. Propolis is obtained by scrapping it from the frames.

Properties

It contains resins and balsams @55 per cent, ethanol and scented oils @10 per cent and pollen @5 per cent.

Uses

It is used in preparing ointments for treating cuts, wounds and abscesses in cattle. Mixed with vasaline to soothen burns.

e) Royal Jelly

Royal jelly is secreted by gland of nurse bees of the age of 6 to 12 days when the glands are fully active. It is very nutritious food and is fed to the young worker larvae and the queen larva and adult. Royal jelly is milky or light pale in colour.

Properties

It contains proteins @15 to 18 per cent. Proteins are mainly amino acids (alanine, arginine, aspartic acid, gultonic acid, glycine, isoleucine, lysine, methionine, phenyl alanine, tryptophane, tyrosine and serine). It also contains lipids @2 to 6 per cent, carbohydrates @ 9 to 18 per cent (glucose, fructose, melibiose, trehalose, maltose and sucrose) and ash @ 0.7 to 1.2 per cent. Vitamin A, B and C, iron, copper, phosphorus, silicon and sulphur are also present.

Production

The queen cell is trimmed to the level of the royal jelly. After 2 or 3 days of grafting, larvae are gently removed with forceps and the royal jelly is removed with royal jelly spoon. This is stored in refrigerated conditions. In case of *Apis mellifera,* 200mg of royal jelly is obtained from a queen cell.

Uses

Responsible for queen determination. Very nutritious food for human beings.

f) Pollen

It is collected by pollen trap from ingoing pollen foragers. Pollen is a rich protein source for human diet.

Uses

In addition to providing man with very valuable materials as honey and beewax, the honey bees are also useful in aiding in pollination of many of the crops. In fact, it has been claimed that the value of bees in pollination of crops is 10 to 20 times the value of honey and wax they produce. Certain crops like apples, alfalfa and clover almost entirely depend upon bees for their pollination. Even among some regularly self-pollinated crops, the yield is considerably increased after visit of bees.

2. SERICULTURE

Sericulture is an agro-business, it combines agriculture and industry to produce silk as a final product. It is a labour intensive activity that employs a significant number of people directly and indirectly. India holds the distinction of being the only country to produce all five commercial silks viz., mulberry, tropical tasar, oak tasar, eri and muga, the latter of which is unique to India due to its golden yellow shine.

Mulberry sericulture is mostly practiced in the major silk producing states of Karnataka, Andhra Pradesh, Assam and Bodoland (Kokrajhar, Chirang, Baksa, and Udalguri districts of Assam), West Bengal, Jharkhand and Tamil Nadu. Mulberry, Oak Tasar, Muga, and Eri silk are all produced in the North East, making it the only region in the world to do so. The north east region accounts for 18% of India's total silk production. India produces about 14% of the world's raw silk with mulberry silk accounting for more than half of it, largely produced in Karnataka. Silk is exported in many forms, in addition to supplying the country's ever increasing domestic demand and this sector is one of the country's major foreign exchange earners.

Silks of India

Species of Silkworms which in turn feed on a number of food plants. Except for mulberry, other varieties of silks are generally termed non-mulberry silks. The non-mulberry silk is recently christened "Vanya Silk", due to its wild nature. India has the unique distinction of producing all the commercial varieties of silk. There are five major types of silk of commercial importance are detailed below.

a) Mulberry silk

The bulk of the commercial silk produced in the world comes from this variety and is often referred to as mulberry silk. Mulberry silk comes from the silkworm, *Bombyx mori* L. which solely feeds on the leaves of the mulberry plant. These silkworms are completely domesticated and reared indoors. The mulberry sector continues to be predominantly rural and small farmer based with post cocoon activities in the cottage and small industry sectors. Mulberry silk contributes to around 80% of the silk production. In India, the major mulberry silk producing states are Karnataka, Andhra Pradesh, West Bengal, Tamil Nadu and Jammu & Kashmir which together contribute 97% of countries total mulberry raw silk production.

b) Tasar

Tasar (Tussah) is a copperish beige colour, coarse silk mainly used for furnishings and interiors. It is less lustrous than mulberry silk but has its own feel and appeal. Tasar silk is generated by the silkworm, *Antheraea mylitta* which mainly thrives on the food plants of Asan (*Terminalia tomentosa*) and Arjun (*Terminalia arjuna*). The rearings are conducted outdoor in nature on the trees. In India, tasar silk is mainly produced in the states of Jharkhand, Chhattisgarh, and Odisha, besides Madhya Pradesh, Maharashtra, Bihar, West Bengal and Andhra Pradesh. Tasar culture is the mainstay for many tribal communities in India.

c) Oak Tasar

It is a finer variety of tasar generated by the silkworm, *Antheraea proyeli* in India which feed on natural food plants of oak (*Quercus*), found in abundance in the Sub-Himalayan belt of India covering the States of Himachal Pradesh, Jammu & Kashmir, Utarakhand, Assam, Mizoram and Manipur. China is the major producer of oak tasar in the world and this comes from the silkworm *Antheraea perni.*

d) Eri

It is also known as Endi or Errandi. Eri is multivoltine silk spun from open ended cocoons, unlike other varieties of silk. Eri silk is the product of the semi-domesticated silkworm, *Philosamia ricini* that feeds mainly on castor leaves. Eri Silkworm being polyphagous has a wide range of food plants such as tapioca, papaya, payam, kessaru, barkessuru, etc. Eri-culture is a household activity practiced mainly in north eastern region for protein-rich pupae, a delicacy for the tribals in the region. Resultantly, the eri cocoons are open-mouthed and are spun. The silk is used indigenously for the preparation of *chaddars* (wraps) for own use by the tribals. Eri silk fabric is a boon for those who practice absolute non-violence and do not use any product obtained by killing any living creature. Eri silk is now popularized as "Ahimsa Silk". Now Eri silk is getting popular the world over due to its isothermal properties which make it suitable for shawls, jackets and blankets. Besides North-Eastern States, it is also getting popularized in Bihar, West Bengal, Odisha, Uttar Pradesh and Andhra Pradesh. Eri silk is suitable for knit products, underwear, kids wear, denim and other fashion garments.

e) Muga

This golden yellow color silk is a prerogative of India and the pride of Assam State. It is obtained from the wild multivoltine silkworm, *Antheraea assamensis*. These silkworms feed on the aromatic leaves of Som and Soalu plants and are reared outdoor on trees similar to that of tasar. This fabric is one of the world's treasures of fine silk fabrics, woven on foot-powered, hand-operated looms, which creates a subtle unevenness. The natural shimmery golden colour of this rare, wild silk needs no dye to enhance its exquisite beauty. It is a high-value product used in products like sarees, mekhalas, chaddars, etc. Muga culture is specific to the State of Assam and an integral part of the tradition and culture of that State. However, the muga culture is getting popularized in other states viz., West Bengal, Meghalaya and Nagaland due to the availability of Som and Soalu plants.

Silk Production

Silk manufacture is divided into two stages: pre-cocoon and post-cocoon. Sericulturists plant mulberry trees and raise disease-free silkworm seeds that grow on mulberry leaves in the first phase. Because the quality of silk depends on the leaves fed to the silkworms, great care is taken with the initial seeds and mulberry plantation. Silkworms go through four phases of development, from egg to moth and during the caterpillar stage, they form a cocoon around themselves. The silkworm's spun thread is essentially made up of two proteins: fibroin, the fiber and sericin, the glue between the cocoon and the fiber. In the post-cocoon period, reeling occurs. Silk is reeled from cocoons within fifteen days of spinning to achieve the highest grade silk. With 10 kg of cocoons, one roll of silk can be made. Weavers weave the reeled silk into final items for the consumer market. For centuries, reeling of silk threads is being done on reeling machines installed in homes of silk industrialists. A simple process of reeling involves cooking the cocoons in hot water and while the cocoons are still hot, the silk fibers are held with fingers and reeled into threads. With the advent of automation on all fronts of the industry, there is now high capacity automated reeling machines with high productivity and yield with minimum human intervention.

Silkworm rearing

a) Rearing House: A separate house is ideal for rearing silkworm. The rearing house should have a sufficient number of windows to permit cross ventilation. Provision should be made to make it airtight for proper disinfection. The rearing house has to be built in such a way as to provide an optimum temperature of 26-28°C and RH of 60-70% for the growth of silkworms at minimum operational cost.

b) Features: A rearing house should be built depending on the brushing capacity and the method of rearing. The rearing area of 2 sq.ft/DFL for floor rearing and 3 sq. ft/ DFL for shoot rearing is the general criteria. The rearing house should have a main rearing hall, an anteroom (8 x 8 ft), and a leaf preservation room. Maintaining a separate chawki room is ideal. Rearing house should face east-west direction. Rearing house should have facilities to maintain the required environmental conditions. Growing trees around the rearing house help to maintain a favourable environment. Rearing house should be constructed taking into consideration the following points such as effective disinfection, washable floor, etc. 480 sq. ft area is required for rearing 100 DFL.

c) Preparation of Rearing House: The rearing room is to be kept ready after disinfection atleast 3-4 days in advance of the commencement of rearing. Preconditioning of the rearing house is essential *ie.,* arrangement of rearing appliances and provision of essential environmental conditions one day in advance.

d) Preparation for brushing: Before commencement of each rearing, the rearing equipment and rearing houses must be thoroughly washed and disinfected with chlorine dioxide. Chlorine dioxide is sprayed on equipment, walls, roofs and floor uniformly to destroy the disease causing organisms. The rooms should be kept closed for about 24 h after disinfection. The doors and windows should be kept open at least 24 h before the commencement of rearing to avoid traces of disinfectants. To disinfect the rearing room and rearing appliances, chlorine dioxide can be used. 500 ml of chlorine dioxide is mixed with 50 g of activator and this is dissolved in 20 litres of water. To this, 100 g of lime powder has to be mixed.

e) Disinfection: Spray 2 % formalin with 0.3% slaked lime or 2.5 % chlorine dioxide with 0.5 % slaked lime @2 litre/m^2 areas for disinfecting the rearing house immediately after completion of rearing and three days before brushing. Dip the rearing equipment in 2% bleaching powder solution and sun dry before use. Dust

5% bleaching powder with slaked lime powder @ 200 g/m^2 around the rearing house and passages and sprinkle water @1 litre/m^2 floor area.

f) Incubation of egg and hatching: The egg sheets should be spread out as a single layer in a chawki tray. The temperature of 25°C and humidity of 80 percent are maintained. For this, paraffin papers and wet foam pads may be used. When the eggs come to the head pigmentation stage (about 48 hours before hatching), they should be kept in dark condition by wrapping them in black paper or by keeping them in a box (black boxing). On the expected day of hatching, eggs are exposed to light, early in the morning to ensure uniform hatching. This facilitates the uniform development of the embryo. Most of the eggs (90 to 95 percent) will hatch in about 2 to 3 h.

g) Low-cost method of preservation of eggs: The eggs can be kept in an earthen incubation chamber.

h) Brushing: The hatched larvae should not be starved and they must be brushed on a paraffin paper in a rearing tray or blue polythene sheet (Rearing bed). This is done by sprinkling chopped tender mulberry leaves of size 0.5 to 1 cm^2 over the hatched larvae. The larvae crawl onto the leaves. After 8 to 10 min the egg sheet is inverted over the rearing tray and gently tapped. Worms that are still attached to the egg sheets should be gently removed from the tray with a feather. A rearing bed is prepared and some more chopped leaves, if necessary, are sprinkled. To prevent drying of leaves and to maintain the required humidity in the rearing bed, wet foam pads and paraffin paper covering are provided.

i) Young age silkworm (Chawki) **rearing:** In a tray of 120 cm x 90 cm x 105 cm size, 20 DFL are brushed and reared up to the second stage.

j) Selection of leaves: From brushing to the end of the second age, the larvae are fed with tender leaves. The leaves are selected from the largest glossy leaf, 3rd or 4th from the top. The next 6 to 8 leaves are used to rear the young age worms up to II moult. The size of the chopped leaf is around 0.5 to 1.0 sq.cm during 2nd age.

k) Leaf preservation: Silkworm grows best when fed with succulent leaves which are rich in nutrients and moisture. The leaves if not preserved properly, dry up and become unsuitable for feeding. The harvested leaves must be preserved in the fresh condition in a wet gunny cloth. If the climate is too hot and dry, the leaves are preserved in a leaf chamber which is lined with gunny cloth. The cloth is kept wet by spraying water at frequent intervals.

l) Cleaning: It is the process of removing the silkworm excreta and leftover leaves in the rearing bed. In the first age, one cleaning is given just a day before the worms settle for moulting. In the second age, two cleanings are given, one after resuming feeding and the other before the second moult. A net with a mesh size of 0.5 x 0.5 cm is spread over the rearing bed and feeding is given. The worms crawl through the net and come to fresh leaves. The net along with the worms and leaves are transferred to another tray. The leftover leaves and litter are discarded.

m) Moulting: At the time of moulting, care should be taken not to disturb the worms. Correct detection of moult and stopping or resuming feeds are very important for the uniform growth of silkworms. During moult, the rearing bed should be kept thin and dry by applying lime @ 30-50 g/m^2 and should have proper aeration. The third, fourth, and fifth instar larvae are considered late age worms. They are reared in bamboo trays. Newspapers are spread over the trays to absorb excess moisture in leaves and fecal pellets. The temperature and humidity requirement gradually comes down as the stage develops. Leaves of medium maturity are fed in the third and fourth age and coarse leaves are fed in the fifth age. Over matured and yellow leaves should be rejected, since they may induce disease outbreaks.

n) Bed disinfectants: Apply bed disinfectants like TNAU seridust, resham Jyothi, vijetha or sajeevini @ 4 kg/100 dfls.

Stage (Before feeding)	**Bed disinfectant (Qty/100 dfls) (g)**
After 1st moult	50
After 2nd moult	150
After 3rd moult	800
After 4th moult	1000
On fourth day of final instar	2000
Total	**4000**

o) Moulting: Remove the paraffin papers and evenly spread the larvae in the rearing bed 6-8 h before settling for a moult. Provide air circulation to avoid excess humidity inside the room. Provide charcoal stove/heaters to raise the room temperature during winter. Apply lime powder 60 min before the resumption of feeding daily during rainy/winter seasons to reduce the dampness in bamboo trays.

p) Mounting: Apply Sampoorna @ 20 ml (dissolved in 4 l of water) per 100 dfls over the leaves for early and uniform spinning of cocoons. After attaining full growth in the final instar, the worms cease to feed and are ready to spin. Such worms are slightly translucent and raise their heads to find a place for spinning. These worms have to be picked up and transferred to a mountage for spinning cocoons. Mounting of worms should not be delayed as the ripened worms will waste silk. About 800-900 worms/m^2 are to be kept on a mountage. For 100 dfls, about 30 to 40 chandrakis are required. Mountages should be kept under shade in well ventilated place.

q) Care during Spinning: Quality of silk depends on the care taken at the time of spinning. Mature worms are sensitive to temperature, humidity, light, etc., at the time of spinning. The ripe worm requires space equal to the area to the square of the length of its body for spinning. Proper spacing avoids wastage of silk for forming preliminary web and avoids double cocoons. To prevent staining of cocoons, keep mountage in an inclined position so that the urine may drop to the ground.

r) Maintenance of humidity: Fluctuation of humidity causes abrupt thinning and thickening of silk filament. A relative humidity of 60-70% is ideal for spinning. Provide proper ventilation and straw mats below the mountage for excreta. Provide even and moderate lighting. Improper lighting (bright light or dark shadow) causes the crowding of larvae to the shaded areas leading to double cocoons. Remove dead worms and non-spinners on the 2nd day of spinning. To protect the silkworm from predatory ants, apply malathion 5% dust /Lakshman Rekha at the base of mountage stand.

s) Harvesting: The silkworms complete spinning in 2 to 3 days but the cocoons should not be harvested at this time as the worms inside are still in the prepupal stage. Harvesting should be done on the fifth day (7th day for bivoltine hybrids) when pupae are fully formed and hard. Do not harvest when the pupa is in amber colour. Dead and diseased worms on the mountages should be rcmoved before harvest. Marketing of cocoons should be done on the sixth day (8th day for bivoltine hybrids).

- **Shoot rearing for late age worms**: Silkworm larvae consume 85% of their food requirement during the fifth instar. Fifty percent of the labor input is utilized during the last seven days of rearing.
- **Rearing house**: Provide a separate rearing house for shoot rearing in shady areas. A separate room should be provided for young age worm rearing, leaf storing and a hall for late age worm rearing.
- **Shoot rearing rack**: A rearing rack of 1.2m x 11.0 m size is sufficient to rear 50 dfls. Provide a 15 cm border on all sides of the shelf to prevent the migration of the larvae. Arrange the shelves in a three-tier system with 50 cm space between the tiers. Fabricate the rack stand with wood or steel and the rearing seat with wire mesh or bamboo mat.
- **Shoot harvesting**: Harvest the shoots at 1 m height from ground level for 60 to 70 days after pruning. Store the shoots vertically upwards in a dark

cooler room. Provide a thin layer of water (3 cm) in one corner of the storage room and place the cut shoots in the water for moisture retention.

- **Feeding**: Provide a layer of newspaper on the rearing shelf. Disinfect the bed and spread the shoot perpendicular to the width of the bed. Place top and bottom ends of the shoots alternatively to ensure equal mixing of different qualities of leaves. Transfer the third instar larvae to shoots immediately after moulting. Watch for feeding rate from the 4th day of the fourth instar. If 90% of larvae have not settled for moulting, provide one or two extra feedings. Provide 3 feedings during rainy/winter months and 4 feedings during summer rearing.
- **Spacing**: 18-36 m^2/100 dfls.

t) Bed cleaning: Bed cleaning is done once during the second day of the fifth instar following the rope (or) net method. In the rope method, spread 2 m length of rope (two numbers) at parallel row leaving 0.5m on another side. After 2 to 3 feedings, the ends of the ropes are pulled to the center to make it into a bundle. In the net cleaning method, spread 1.5 cm^2 size net across the bed. After 2 or 3 feedings, the nets are lifted and the old bed is cleaned and disinfected. Transfer the net to a newer shelf and spread the net over the shoots; larvae will migrate to the lower layer.

Advantages

- Labour saving upto 70% when compared on an hour-to-hour basis with the leaf feeding method. Leaf saving up to 15-20%. Hence, the leaf cocoon ratio is less by 2-3 kg and extra cocoon production.
- Better cocoon characters and effective rate of rearing (ERR). Better preservation of leaf quality both during storing and on the bed.
- More organic matter production (up to 18 tonnes per ha per year).
- Better hygienic conditions can be maintained. Handling of silkworms minimized. Hence, contamination and spreading of disease are reduced.
- Bed cleaning only once after IV moult. Worms and leaves are kept away from the litter. Hence, the chances of secondary contamination are minimized.
- Labour dependent risk is reduced.

Disadvantages

- Required more rearing room floor area (by 30%).
- Bed refusals will not be available as cattle feed.
- Planting materials (cuttings) will not be available.

u) Time of harvest: In sericulture, harvest signifies the collection and gathering of produced cocoons from a spinning tray or mountage. For completion of spinning and metamorphosis, about 5-6 days are required. While harvesting, it may also be considered during summer, the process is faster, whereas in cooler months, it is slower. After pupation, when the integument of the pupa turns brown and hard on 5th day, the cocoons may be harvested. The safest method is to check the condition of pupae by slit open a few pupae. Premature collection of cocoons or harvesting may lead to loss of silk content of the cocoons due to incomplete spinning or killing the delicate pre-pupa or pupa within the cocoons. To avoid this, harvest cocoons on 6th day after the completion of the full 5 days in mountage, counted from the last day of mounting. The pupa will be hard and cocoon shell will be dry. This condition will be suitable for safe handling and transportation of cocoon. Cocoons release moisture even after completion of spinning. Delayed harvest will lead to weight loss of cocoons.

- **Method of harvest:** Method of harvesting of silk cocoons varies depending upon the mountage (spinning tray) used. Whatever the method of harvest, first remove litters and left over of leaves, dead or un-spun larva, naked pupa (without cocoon), flimsy and melted cocoons from the mountage. Flimsy and melted cocoons may spoil the good cocoons by spilling stain. Spiral bamboo mountage (chandrika) is most commonly used in India. Harvesting is done mostly by hand picking. In case of plastic collapsible mountages, harvest manually by hand picking. In the rotary card board mountages, harvesting by hand is rather difficult and time consuming. Wooden harvester can be used for quick harvesting. It has two parts. First one is a wooden frame on which cardboard mountage can be fixed. Another is pusher, made up of wooden pegs, that fits with the hole of mountages. After removing dead worms, flimsy and stained cocoons, fix cardboard mountage on the harvester. Then, place the wooden pusher on the holes and push gently so that cocoons come out of cardboard holes and collect the cocoons by hand by folding the mountages. In order to ease the process, machines are developed by CSR&TI, Mysore for harvesting and deflossing the cocoons which are detailed below.
- **Cocoon harvester**: It is made up of iron or wood. It consists of two parts and it is used for harvesting cocoons from rotary mountages. Cardboard mountages are inserted into the wooden frame and cocoons are pushed out using the pusher made of the same size as the holes of the mountage.
- **Cocoon deflosser**: Hand operated and motorized cocoon deflossing machines of different capacities are available for fast deflossing of the cocoons. The motorized machines have the capacity to defloss100 kg cocoons per hour.

After harvesting, spread the cocoons in trays with a maximum of two layers of the cocoon. Heaping of cocoons may lead to accumulation of moisture and heat and ultimately melting of cocoons.

- **Deflossing:** After harvesting cocoons, the cocoons are covered with a soft and loose layer of silk filaments, which is called floss. Removing the flossy layer from the cocoon is called Deflossing. Deflossing is a prerequisite for the extraction of silk (reeling). During the process of reeling, floss is removed by brushing. Presently, many manuals or power operated deflossing machines are available. These machines are quite efficient for all kinds of cocoons. The cost of deflossing by machines is about 40-50 times cheaper than manual. In the market, deflossed cocoons will fetch a higher price as the buyer can assess the quality and silk content very quickly.
- **Sorting of cocoons:** Removal of odd shaped and defective cocoons from the lots is called sorting of cocoons. Defective cocoons affect the reeling performance and quality of the silk. Therefore, unsuitable cocoons are to be sorted out from the good cocoons to get optimum results during reeling. The cocoons are sorted into a) good b) double c) uzi pierced d) flimsy or thin-shelled e) melted or stained and f) thin-end cocoons. Sorting will improve the product image and marketability. Reelers always prefer uniform shapes. Uniform shape improves reeling performance and silk quality. Sorted out defective cocoons can be sold at different rates.
- **Assessment of cocoons:** Cocoons produced are to be marketed. Following points/ characters are generally considered to assess the cocoons in fixing the price.
- **Cocoon Weight**: Cocoons are being sold on a weight basis. The weight of the cocoons gradually decreases due to moisture loss and consumption of the fat till the pupa transforms into moths and emerges. So, immediately after harvest, defloss, sort and weigh the good cocoons. This will be the weight of green cocoons. Make a label indicating race or combination, date of spinning, green weight, etc. while taking it to the market.
- **Assessment of Cccoons for defective cocoon percentage**: One kg of cocoons from the lot are taken at random. Sort out defective cocoons like double, flimsy, melted, pierced, etc. Count and weigh good and defective cocoons separately.
- **Rendita**: It represents the quantity of green (fresh) cocoons required to produce 1 kg of raw silk. If 10 kg of the green (fresh) cocoon of a lot is required to produce 1 kg raw silk, then the Rendita of that lot is 10. It is an indicator of price fixation of the cocoon when purchased for reeling.

➢ **Shell ratio (%)**: It indicates the quantity of silk shell in terms of cocoon weight and expressed in percentage. This value gives a clear indication of the quantity of raw silk that can be reeled from a lot of fresh cocoons. It is used for estimating Rendita and ultimately helps in price fixation.

Marketing of cocoons

There are separate markets for different types of cocoons like seed cocoons and reeling cocoons. There are many established Government controlled markets in cocoon producing areas where market authorities first check the cocoons for complete pupation and then arrange for display and bidding of cocoons and as a producer, then the products will sell to the highest bidder. In that way, the right price for cocoons ca be obtained.

Processing of mulberry cocoons

Quality of raw material, reeling machinery, reeling process parameters, the human skill involved in manual and mechanical operations and quality of water are the major factors that have a direct bearing on the productivity and quality of raw silk. Among these, quality of cocoons plays a major role in raw silk yield and productivity. In order to achieve better reeling performance and quality raw silk, the role of reeling technology is significant.

i) Drying of cocoons

Cocoons should be hot air-dried to the optimum level to achieve better quality silk. This method achieves the major objective of hardening the sericin uniformly in all the layers of the cocoon shell apart from killing pupa and removing the moisture. Hot air drying can be done by batch-type hot air driers or in Ushnakoti. In case the cocoons are not dried to the optimum level, then may be once again dried at lower temperature *viz.,* 70 °C and 60 °C depending upon the requirement.

Temperature	Duration
115°C	60 minutes
100°C	60 minutes
85°C	60 minutes
70°C	60 minutes
55°C	60 minutes

ii) Cocoon storage

The cocoons dried to the optimum level may be conditioned for a minimum period of 7-10 days before taking up for reeling in order to achieve better reeling performance. The cocoons should be stored in an appropriate storage room for long duration storage. The cocoon storage should have 20°C and below the temperature in the center of the room and the air inside the room should have 55 % and below relative humidity, so that fungus will not attack the cocoons.

iii) Cocoon sorting

In order to achieve better results in reeling, the defective cocoons namely, double cocoons, flimsy cocoons, melted cocoons, uzi infected cocoons, urinated cocoons and malformed cocoons which are unsuitable for reeling quality silk should be sorted out using CSTRI cocoon sorting table.

iv) Cocoon mixing

Mixing of cocoons should be done based on the single cocoon filament denier and reelability of the cocoon lots. Among the cocoon lots taken for mixing, cocoon lots having similar single cocoon filament denier and reelability should be taken. The cocoon lots having a wide variation in single cocoon filament denier and reelability should not be mixed together.

v) Cocoon cooking

The ideal cocoon cooking for bivoltine cocoons is done by employing two/ three pan cooking and circular type pressurized cocoon-cooking machine. In order to cook the hot air dried bivoltine cocoons effectively, three-pan cooking can be adopted. The procedure while three pan cooking (with water as media) is as follows.

- The cage with cocoons is immersed in the first cooking pan at 50 -55°C for about 45 - 60 seconds.
- The cage is transferred to the second cooking pan, which is at 90 - 93°C for about 90-120 seconds.
- The cage is then dipped in the third pan at 60 - 65 °C for about 45-60 seconds.
- The cage is then returned to the second pan at a boiling temperature of about 95-96°C for 90-120 seconds.
- Steam supply is stopped and cold water is sprinkled on the cage for about 10 seconds.

- The cage is opened and the cocoons are transferred to the cooking pan. The floating cocoons are then pressed using a perforated pressing plate for 30 seconds without steam and 30 - 45 seconds with steam.
- Steam supply to the pan is stopped and cold water is sprinkled gradually on the cocoons to reduce the temperature to 70 - 80 °C gradually.
- The cocoons are then brushed at 75 - 80°C and transferred to a reeling basin for large-scale reeling establishment. It is advisable to adopt a circular pressurized cooking technique for hot air dried bivoltine cocoons to achieve better productivity and quality raw silk.

Value addition

Proper utilization of secondary and waste products of sericulture industry can generate extra income in addition to the silk. The major wastes and by products of sericulture are sericin, pupae, moths, silkworm excreta, silk fiber waste and mulberry leaves, fruits and roots. The new commercial products can be obtained from these raw materials with the valuable destination for pharmaceutical (anti diabetic, antiviral, hypoglycemic, antibacterial and antivirus products), cosmetics (skin and hair products), zootechnicals (fodder for caws, rabbits, goats, poultry, pigs, sheep and fur animals), foodstuff (oil, juice, marmalade, wine, fruit distillate, vinegar, dried fruit powder, natural colouring) and ecological use (afforestation and phytoremediation).

a) Mulberry

- **Mulberry leaves:** In addition to being mainly used as feed for silkworm, the leaves can be utilized for cattle feeding, preparation of chlorophyll, soap, cosmetics, hair oil and mulberry tea. Leaves also hold pharmaceutical value as they have diuretic and anti-diabetic properties.
- **Mulberry stem:** It is used as a fuel wood and in the preparation of sports goods, toys, fine quality papers and collection of edible fungus. Tannins extracted from stem can be used as a coloring agent.
- **Mulberry fruits**: It is used for the preparation of juice, jelly, squash and jam.
- **Silkworm litter:** It is used in compost making, biogas production and also used as cattle and poultry feed. Chlorophyll paste, proteins and plastic materials can be prepared.
- **Silkworm faeces:** It is generally considered as waste. It is mostly used in compost / vermin compost making as cattle feed. It can be used as aqua feed and for preparation of chlorophyll paste.

- **Pupae:** Silkworm cocoon consists of two major portions *viz.,* cocoon shell (12.24%) and pupae (76 – 88%). Dried pupae contain 28.4% oil and 51.3% proteins, which are used for the manufacture of pupal oil, amino acid powder, peptides / protein concentrate. The pupal oil which is brown in colour with bad odour finds limited use in the manufacture of industrial soap. It finds use in the preparation of cosmetics, lotions, cream and shampoo, etc. Two pupal oil extract units are functioning at Mysore and Kallur (Andhra Pradesh). The pupal waste after extraction of oil can be used as poultry / aqua feed. Pupal fat is a good raw material for the manufacture of glycerin soap and cosmetics.
- **Pupal skin:** It is available in the reeling and grainage as waste. It is made up of chitin (3-4 % of dry weight of pupae). It is used as an additive to increase the loaf volume of wheat bread.
- **Basin Refuse:** The last layer of cocoon attached to pupae (Pelade layer) cannot be reeled. This can be used for manufacture of Spun silk which will reduce the cleaning cost of pupal wastes.
- **Cut and pierced cocoons:** Cut and pierced cocoons can be used for the production of hand span yarn (matka) using takli ; such units are functioning in Malda, West Bengal.
- **Noil yarns:** First quality of waste obtained during the processing of spun silk yarn is utilized for production of 'Noil yarn; which are used for the manufacture of carpets.
- **Twisting / throwers wastes:** Waste silk obtained during winding, twisting and doubling of yarn are used for the production of thick fabrics / carpets.

Suggested Readings

Abdul Quium Osmani 2021 *An Introduction to Non--Mulberry Sericulture,* Guwahati Ashok Book Stall

Belsare DK Rakesh Kumar Singh Shashikala Belsare D and Ravi H Deshmukh 2019 *Textbook of Apiculture (Beekeeping),* Belur, West Bengal: Himalaya Publishing House

Johnson M and Kesary M 2019 *Sericulture,* Nagercoil, Tamil Nadu: Saras Publication

Madan Mohan Rao M 2019 *An Introduction to Sericulture,* Hydrabad: BS Publications BSP Books

Tamilselvi M and Abdul Jaffar Ali H 2018 *A Textbook on Apiculture,* Chennai: Vijay Nicole Imprints

CHAPTER 15
Mineral Products

M.P. Divya and M. Packialakshmi
Tamil Nadu Agricultural University, Coimbatore

Mineral products are defined as minerals in raw or processed form, bricks, dressed stone, rock aggregates, chips, ballast, stone dust, sand or any product to be prepared from minerals without involving any chemical changes. The availability of mineral resources determines the socioeconomic development of a country. Mineral resources provide mankind with 95% of the energy it needs and 80% of the industrial raw materials it uses for agricultural production. Minerals or inorganic substances found naturally in the earth's crust are pure inorganic compounds. The number of minerals identified exceeds 2000 and most of them are inorganic, formed from a combination of elements. In addition, a small portion of the earth's crust consists of organic materials, which are composed of single elements like gold, silver, diamonds and sulfur.

Mineral resources, while principally nonrenewable, are generally placed in two main categories: metallic (containing metals) or nonmetallic (containing other useful materials). Metallic minerals are those from which valuable metals (eg. iron, copper) can be extracted for commercial use. Metals that are considered geochemically abundant occur at crustal abundances of 0.1 percent or more (eg. iron, aluminum, manganese, magnesium, titanium). Metals that are considered geochemically scarce occur at crustal abundance of less than 0.1 percent (eg. nickel, copper, zinc, platinum metals). Some important metallic minerals are hematite (a source of iron), bauxite (a source of aluminum), sphalerite (a source of zinc), and galena (a source of lead). Metallic minerals occasionally but rarely occur as a single element.

Types of minerals

a) Metallic minerals

Metals are obtained by the extraction of these minerals. These are very valuable as they provide metals in pure form.

- **Aluminum** is the most abundant metallic element in the earth's crust. Bauxite ore is the main source of aluminum. Aluminum is used in automobiles and airplanes (36%), bottling and canning industries (25%), building and electrical (14%) and in other applications (25%).

- **Basalt** is an extrusive igneous rock. Crushed basalt is used for railroad ballast and aggregate in highway construction and is a major component of asphalt.
- **Barium** is an element, derived primarily from the mineral barite and used as a heavy additive in oil-well-drilling mud, paints, rubber, plastic and paper, production of barium chemicals and glass manufacturing.
- **Beryllium** is an element commonly associated with igneous rocks, has industrial and nuclear defense applications and is used in the light, very strong alloys for the aircraft industry. Beryllium salts are used in x-ray tubes and as a deoxidizer in bronze metallurgy. The gemstones of beryl, a beryllium mineral are emerald and aquamarine.
- **Bismuth** is used in different applications. The majority is consumed in bismuth alloys, pharmaceuticals and chemicals. The remainder is used in ceramics, paints, catalysts and a variety of minor applications. Bismuth metal is relatively inert and non-toxic. It has replaced toxic lead in many applications such as plumbing, bullets, birdshot, metal alloys, and soldering. Bismuth compounds are used in stomach-upset medicines, treatment of stomach ulcers, soothing creams and cosmetics.
- **Boron** compounds are used for different purposes in industry and the home. Boron is used to make glass, ceramics, enamels, fiber glass, water softeners, soaps and detergents. Other uses are in agricultural chemicals, pest control, fire retardants, fireworks, medicine and various minor applications. Boron nitride is one of the hardest known substances and is used for abrasives and cutting tools.
- **Bromine** is recovered commercially through the treatment of seawater brines and is used in leaded gasoline, fire extinguishers and retardants, well-completion fluids and sanitary preparations. Bromine is the only liquid nonmetallic element.
- **Calcium** occurs as calcium carbonate. It is used in adhesives and sealants, cosmetics, food, paint, paper, pharmaceuticals, plastics and rubber for the production of lime and as crushed stone in construction. Immense quantities of calcium are found in sedimentary rock deposits of gypsum, limestone and shale. Some common calcium bearing minerals include apatite (calcium phosphate), calcite (calcium carbonate), dolomite (magnesium carbonate), fluorite (calcium fluoride), and gypsum (calcium sulfate). Calcium metal is produced in Canada, China, France, Russia and the United States. Total world output is thought to be less than 6,000 metric tons per year. United States

consumption of calcium metal is small. On a worldwide basis, more than 100 million metric tons per year of apatite and gypsum are mined and calcite and dolomite are produced in billions of metric tonnes per year.

- **Cadmium** is used in plating and alloying, pigments, plastics and batteries. Cadmium is obtained from the ore minerals sphalerite (Zn, Cd) and Greenockite (Cd).
- **Chromium** is used in the production of stainless and heat-resistant steel, full-alloy steel, super alloys and other alloys. Chromium is obtained from the ore mineral Chromite (Mg, Fe, Cr, Al, Fe_2O_4.)
- **Clays** have different clay minerals which are used for industrial applications. Clays are used in the manufacturing of paper, refractories, rubber, ball clay, dinnerware and pottery, floor and wall tile, sanitary wear, fire clay, firebricks, foundry sands and drilling mud, iron-ore pelletizing, absorbent and filtering materials, construction materials and cosmetics.
- **Cobalt** is used in corrosion and abrasion resistant alloys with steel, nickel and other metals for the production of industrial engines. Other uses of cobalt metal include magnets and cutting tools. Cobalt salts are used to produce a blue color in paint pigments, porcelain, glass and pottery. Cobalt is obtained from the ore minerals Linneite (Co_3S_4), Cobaltite CoAsS, and (Fe, Ni, Co).
- **Copper** is used in electric cables and wires, switches, plumbing, heating, electrical and roofing materials, electronic components, industrial machinery and equipment; transportation, consumer and general products, coins and jewellery.
- **Fluorite** is used in the production of hydrofluoric acid, which is used in the pottery, ceramics, optical, electroplating and plastic industries. It is also used in the metallurgical treatment of bauxite, as a flux in open-hearth steel furnaces and in metal smelting as well as in carbon electrodes, emery wheels, electric arc welders and toothpaste as a source of fluorine.
- **Gold** is used in dentistry and medicine, jewellery and arts, medallions and coins and ingots. It is also used for scientific and electronic instruments, computer circuitry, as an electrolyte in the electroplating industry and many applications in the aerospace industry.
- **Garnet** is used in water filtration, electronic components, ceramics, glass, jewellery and abrasives used in wood furniture and transport manufacturing. Garnet is a common metamorphic mineral that becomes abundant enough to mine in a few rocks.
- **Halite** is used in the human and animal diet as food seasoning and food preservation. It is also used to prepare sodium hydroxide, soda ash, caustic

soda, hydrochloric acid, chlorine and metallic sodium, and it is used in ceramic glazes, metallurgy, curing of hides, mineral waters, soap manufacture, home water softeners, highway deicing, photography and scientific equipment for optical parts.

- **Iron Ore** is used to manufacture steels of various types and other metallurgical products such as magnets, auto parts and catalysts. The earth's crust contains about 5% iron, the fourth most abundant element in the crust.
- **Iodine** is used as an antibacterial agent in soaps and cleaning products in restrooms, in iodized salt to prevent goiter and in first aid boxes as an antiseptic.
- **Magnesium** is used in cement, rubber, paper, insulation, chemicals and fertilizers, animal feed, and pharmaceuticals. Magnesium is obtained from the ore minerals Olivine, Magnesite and Dolomite.
- **Manganese** is essential for iron and steel production. Manganese is obtained from the ore minerals viz., Braunite, Pyrolusite and Psilomelane.
- **Mercury** is extracted from the mineral cinnabar and is used in electrical products, electrolytic production of chlorine and caustic soda, paint and industrial and control instruments (thermometers and thermostats).
- **Silver** is used in photography, chemistry, electrical and electronic products, fine silverware, electroplated wire, jewellery, coins and brazing alloys and solders.
- **Sulfur** is of importance to every sector of the world's manufacturing processes, drugs and fertilizer complexes. Sulfur is used as an industrial raw material through its major derivative, sulfuric acid. Sulfuric acid production is the major end use for sulfur. Most sulfur goes into fertilizer; oil refining is another major use as well as a source of sulfur.
- **Granite** can be cut into large blocks and used as a building stone. When polished, it is used for monuments, headstones, countertops, statues and facing on buildings. It is also suitable for railroad ballast and road aggregate in highway construction.
- **Graphite** is the crystal form of carbon. Graphite is used as a dry lubricant and steel hardener for brake linings and the production of "lead" in pencils. Most graphite production comes from Korea, India and Mexico.
- **Pyrite** is used in the manufacture of sulfur, sulfuric acid and sulfur dioxide; pellets of pressed pyrite dust are used to recover iron, gold, copper, cobalt and nickel.

- **Quartz** crystals are popular as a semiprecious gemstone; crystalline varieties include amethyst, citrine, rose quartz and smoky quartz. Because of its piezoelectric properties, quartz is used for pressure gauges, oscillators, resonators, and wave stabilizers. Quartz is also used in the manufacture of glass, paints, abrasives, refractories and precision instruments.
- **Molybdenum** is used in stainless steels (21%), tool steels (9%), cast irons (7%), chemical lubricants (8%) and in other applications (55%). It is commonly used to make automotive parts, construction equipment and gas transmission pipes and pure metal molybdenum is used as filament supports in light bulbs, metalworking dies and furnace parts because of its high melting temperature.
- **Nickel** is vital as an alloy to stainless steel, and it plays a key role in the chemical and aerospace industries. Leading producers are Canada, Norway and Russia.
- **Platinum Group Metals (PGMs)** include platinum, palladium, rhodium, iridium, osmium and ruthenium. These elements commonly occur together in nature and are among the scarcest of the metallic elements. Platinum is used principally in catalytic converters for the control of automobile and industrial plant emissions; in jewellery; in catalysts to produce acids, organic chemicals and pharmaceuticals and in dental alloys used for making crowns and bridges.
- **Potash** refers to a group of water-soluble salts containing the element potassium, as well as ores containing these salts. Potash is used in fertilizer, medicine and the chemical industry and to produce decorative colour effects on brass, bronze and nickel.
- **Tin** is used in the manufacture of cans and containers, electrical equipment, and chemicals.
- **Titanium** is a metal used mostly in jet engines, airframes and space and missile applications. In powdered form, titanium is used as a white pigment for paints, paper, plastics, rubber and other materials.

b) Non-Metallic Minerals

The minerals which do not contain metals are called non-metallic minerals. Examples of non-metallic minerals are Diamond, Mica, Salt, Potash, etc. The Kohinoor diamond placed on the crown of England's queen is an example of a non-metallic mineral.

- **Mica** is a naturally occurring non-metallic mineral that is based on a collection of silicates. Mica is a very good insulator that has a wide range of applications in the electrical and electronics industries. It can withstand high voltage and has a low power loss factor. It is used in toothpaste and cosmetics because of its glittery appearance. It also acts as a mild abrasive in toothpaste. India is one of the foremost suppliers of mica. Mica-bearing igneous rocks occur in Andhra Pradesh, Bihar, Jharkhand, Maharashtra and Rajasthan. India has a near monopoly in the production of mica (60 % of the world's total production). Production decreased in recent times due to a fall in demand in the international market. The fall in demand is due to better synthetic alternatives that are available.

- **Limestone** rocks are composed of calcium carbonate, the double carbonate of calcium and magnesium, or a mixture of both. Limestone also contains small quantities of silica, alumina, iron oxides, phosphorus and sulphur. 75 percent of limestone is used in the cement industry, 16 percent in the iron and steel industries and 4 percent in the chemical industries. The rest of the limestone is used in paper, sugar, fertilizers, etc. Almost all the states of India produce some quantity of limestone. Over three-fourths of the total limestone of India is produced by Madhya Pradesh, Rajasthan, Andhra Pradesh, Gujarat, Chhattisgarh and Tamil Nadu.
- **Dolomite** has limestone with more than 10 percent of magnesium. When the percentage rises to 45, it is true dolomite. Dolomite is mainly used as blast furnace flux, as a source of magnesium salts and in fertilizer and glass industries. The iron and steel industries are the chief consumer of dolomite (90 percent) followed by fertilizer, ferro alloys, and glass. Dolomite is widely distributed in all parts of the country. Orissa, Chhattisgarh, Andhra Pradesh, Jharkhand, Rajasthan, and Karnataka are the main dolomite producing states and contribute more than 90 percent of the total production. Odisha and Chhattisgarh together account for about 57 percent dolomite of India.
- **Asbestos** has two quite different minerals viz., i) a variety of amphibole and ii) a fibrous variety of serpentine (chrysotile). Chrysotile is an important variety and accounts for 80 percent of the asbestos in commercial use. Asbestos has great commercial value due to its fibrous structure, filaments of high tensile strength and its great resistance to fire. It is widely used for making fire-proof cloth, rope, paper, millboard, sheeting, etc. It is also used in making aprons, gloves, brake-linings in automobiles, etc. Asbestos cement products like sheets, pipes and tiles are used for building purposes. When asbestos is brittle, it is made into filter pads for filtering acids. Mixed with magnesia, it is used for making 'magnesia bricks' used for heat insulation. Rajasthan and Andhra Pradesh produce almost the whole of asbestos in India. Rajasthan is the largest producer. Important occurrences are known

in Udaipur, Dungarpur, Alwar, Ajmer and Pali districts. In Andhra Pradesh, asbestos of fine quality occurs in Pulivendla taluk of Cuddapah district.

- **Magnesite** is an alteration product of dunites (peridotite) and other basic magnesian rocks. It is primarily used for manufacturing refractory bricks. It is also used as a bond in abrasives, manufacture of a special type of cement for artificial stone and tiles and for extraction of the metal magnesium. The steel industry also uses magnesite. Major deposits of magnesite are found in Uttaranchal, Tamil Nadu and Rajasthan. Tamil Nadu is the largest producer of magnesite in India. Tamil Nadu has one of the largest deposits of magnesite in the world and the largest in India are found at Chalk Hills near Salem town.
- **Kyanite** occurs in metamorphic aluminous rocks. It is primarily used in metallurgical, ceramic, refractory, glass and cement industries due to its ability to stand high temperature. It is also used in making sparking plugs in automobiles. India has the largest deposits of kyanite in the world. Kyanite grades depend on aluminium content. Greater the aluminium content, the greater the quality. Jharkhand, Maharashtra and Karnataka produce the whole kyanite of India.
- **Sillimanite** found in Tamil Nadu, Orissa, Kerala, Andhra Pradesh and West Bengal. Orissa is the largest producer of sillimanite in India. Ganjam district is an important sillimanite producing district. Kerala is the second largest producing state.
- **Gypsum** is a hydrated sulphate of calcium. It is a white opaque or transparent mineral. It occurs in sedimentary formations such as limestones, sandstones and shales. It is mainly used in making ammonia sulphate fertilizer and in the cement industry. It makes up to 4-5 per cent of cement. It is also used in making plaster of paris, moulds in ceramic industry, tiles, plastics, etc. It is applied as surface plaster in agriculture for conserving moisture in the soil and for aiding nitrogen absorption. Rajasthan is the largest producer of gypsum in India (99 per cent of the total production of India). The main deposits occur in the tertiary clays and shales of Jodhpur, Nagaur and Bikaner. Jaisalmer, Barmer, Chum, Pali and Ganganagar also have some gypsum bearing rocks. The remaining gypsum is produced by Tamil Nadu, Jammu and Kashmir, Gujarat and Uttar Pradesh. Water and phosphoric acid plants are important sources of by product of gypsum. Marine gypsum is recovered from salt pans during the processing for common salt in Gujarat and Tamil Nadu. Phospho-gypsum is obtained as a byproduct while manufacturing phosphoric acid whereas fluro-gypsum is obtained while manufacturing aluminium flouride

and hydro-fluoric acid. The recovery of by-product viz., phospho-gypsum, fluoro- gypsum and marine gypsum together are higher than mineral gypsum.

- **Salt** is obtained from salt pans in lakes and rocks. Rock salt is taken out in the Mandi district of Himachal Pradesh and in Gujarat. It is less than 1 per cent of the total salt produced in India. Sambhar Lake in Rajasthan produces about 10 per cent of annual production. Sea brine is the source of salt in Gujarat, Maharashtra and Tamil Nadu. Gujarat coast produces nearly half of the salt.

Suggested Readings

Andrey Bulakh and Hans-Rudolf Wenk 2016 *Minerals: Their Constitution and Origin,* Cambridge, England: Cambridge University Press

Dan Binkley and Richard F Fisher 2019 Ecology and Management of Forest Soils, John Wiley & Sons Ltd.440p

Donald L Grebner Peter Bettinger and Jack P Siry 2014 Introduction to forestry and Natural Resource Management, Academic Press496p

Joaquim Sanz Oriol Tomasa Abigail Jimenez-Franco and Nor Sidki-Rius 2022 *Elements and Mineral Resources,* Switzerland AG: Springer Nature

Mahapatra GB 2019 *A Textbook of Geology*, New Delhi: New Delhi: *CBS Publishers &* Distributors Pvt. Ltd

Martin Okrusch and Hartwig E Frimmel 2020 *Mineralogy: An Introduction to Minerals, Rocks, and Mineral Deposits,* Berlin, Heidelberg: Springer

Rupert Hochleitner 2022 *The Minerals Encyclopedia*, Canada: Firefly Books Ltd